Global Environmental Careers

The Worldwide Green Jobs Resource

Justin Taberham CEnv

WILEY Blackwell

Registered Offices
John Wiley & Sons, Inc., 111 River Street, Hoboken, NJ 07030, USA
John Wiley & Sons Ltd, The Atrium, Southern Gate, Chichester, West Sussex, PO19 8SQ, UK

Editorial Office
9600 Garsington Road, Oxford, OX4 2DQ, UK

For details of our global editorial offices, customer services, and more information about Wiley products visit us at www.wiley.com.

Wiley also publishes its books in a variety of electronic formats and by print-on-demand. Some content that appears in standard print versions of this book may not be available in other formats.

Library of Congress Cataloging-in-Publication data applied for

Hardback: 9781119052845

Cover Design: Wiley
Cover Image: Courtesy of Bailey Taberham

Set in 9.5/12.5pt STIXTwoText by Straive, Pondicherry, India
Printed and bound by CPI Group (UK) Ltd, Croydon, CR0 4YY

C9781119052845_090921

I would like to dedicate this book to my Mum who inspired my love for books; my Dad who instilled in me the need to work hard and have 'stickability'; and to Jasmine for her constant love and support. Many thanks to my children for their involvement – Bailey for the book cover image, Ethan for keeping me well fed and amused and Lara for proofreading. This book is in memory of my Dad and Mama Lee, who gave me great support while I was writing but never got to see it finally published.

Contents

Acknowledgements

I would like to thank the entire team at Wiley who have helped to bring this book to fruition and most notably to Andrew Harrison, Senior Commissioning Editor and Mandy Collison, Managing Editor.

A group of reviewers helpfully reviewed the book text and passed on their expertise to help make the book more valuable to readers. They are Carol McClelland Fields, Kevin Doyle, Lisa (Yee) Yee-Litzenberg, Laura Thorne and Sharmila Singh.

I would like to thank my networks and the numerous volunteer contributors, text reviewers and experts globally who have made the book more global, relevant and information rich.

The names of many of the major contributors are noted at the end of the book, but there are also hundreds of additional helpers who chipped in with views and helpful ideas. My thanks go to all.

Justin Taberham, London
June 2021

1

Introduction

1.1 Author Introduction

The author, Justin Taberham, is an environmental professional and Chartered Environmentalist with more than 30 years' experience in the global environment sector, working for Government agencies, a water utility, a lobbying charity and a global professional body. He is now a Consultant and Advisor, working in environmental publications, business and careers. He has significant experience of environmental recruiting, developing staff and advising and mentoring people globally who seek to develop a career in the environment sector. His website is www.justintaberham.com, and the site has a full professional profile.

The Env.Careers Website

The author manages the Env.Careers website www.env.careers which is 'The "one stop shop" for people who want to develop a career in the global environment sector'.

The Env.Careers website is an allied resource to this book, and it helps deliver ongoing additional content. The website is also a standalone resource with tips, advice and mentoring for those seeking a green career.

1.2 The Author's Green Career – Any Lessons to Learn?

I grew up in the Norfolk countryside, in rural England. I was an enthusiastic scout, angler, hill walker and conservation volunteer. I was also keenly involved in environmental campaigns on issues such as Rainforest Conservation, Acid Rain and Species Loss. My parents were very tolerant of me bringing home 'roadkill' and other natural finds to 'study'.

As I got older and had to make career direction and subject choices at school, I was often shunted into a specific area of interest which would enable me to more easily get a 'standard' job – early suggestions from careers questionnaires were Policeman, Insurer and Teacher. None of these were of much interest to me – it was a mystery why the questionnaires never asked, 'What do you actually want to do?' Careers advice tended to just look at the limited subjects I was studying and extrapolate what career field was broadly relevant; the more

Global Environmental Careers: The Worldwide Green Jobs Resource, First Edition. Justin Taberham.
© 2022 John Wiley & Sons Ltd. Published 2022 by John Wiley & Sons Ltd.

sophisticated current career survey techniques still extrapolate from your school subjects, general interests and personality traits and find you very common areas of work, in an almost apologetic attempt to pigeonhole you.

In the early 1980s, very few people worked in the environment sector and I struggle to remember any careers talks that mentioned it. 'The Environment' tended to be a subject for hippies and drop-outs, and environmentalism was not perceived in the more positive light that it is today. The 'Natural World' was being promoted to the public by the likes of David Attenborough and Jacques Cousteau, but it seemed very exotic and separate from normal career options. Protecting and managing the environment as a career wasn't promoted, even following the major nuclear reactor accident at Chernobyl in 1986 and the scourge of acid rain in the 1980s.

Many people I have met who work in the global environment sector have grown their passion from childhood experiences with nature and I'm an example of this. Just because a quality has been developed in early life, this doesn't preclude it from playing a part in your future career decisions.

I decided at school that I would take 'general' A-Levels (UK examinations pre-university), investigate environmental degrees and consider what jobs were on offer. It must be noted that this was a time when internet searches were not possible, and university and jobs information had to be found in a school or public library, by searching the indexes and shelves. School careers services were very limited and only the most 'popular' career sectors had information booklets.

My parents were wonderfully supportive of my decisions and I enrolled in the leading Environmental 'sandwich' degree at what is now the University of Hertfordshire, UK (a sandwich degree is a degree that has a year in 'industry' in the middle). From the first day, I was hooked and realised that this was a career direction worth pursuing with a passion. Because of the sandwich degree format, I gained a year's experience in fisheries management around London and Hertfordshire, which helped me to get on the jobs ladder after graduating, in 1990. However, getting my first job still took four months of odd jobbing, hundreds of applications and endless job hunting until things worked out.

One thing that strikes me, when I look back at my attitude at this time, was my 'stickability' in the light of people giving me advice to go for a mainstream job. In the global environment sector, this trait is hugely important. Throughout this book, the importance of personality and passion will be mentioned repeatedly, and I am unapologetic for comments that are made frequently when they are crucial.

To complete a degree is a great help for certain roles in the environment sector, but there are many jobs and organisations where your actual skills and personality count just as much as a higher-level education. There is a significant move away from employers listing specific university qualifications towards the top of their adverts, and instead, professional skills and vocational qualifications are being actively listed. These skills might include specific job-related safety and training courses, IT suite and other certifications. This is of relevance to employers who need to hire people with the professional skills they require, rather than just an academic qualification; it also can reduce the initial training costs for a new member of staff. In a commercial world with tight budgets, a candidate who has already completed some level of training is a plus. Many companies have recruitment sections that explain apprenticeships, internships, openings for 'craft' and manual professionals and school recruitment.

There is also a Masters/PhD dilemma for some people in the sector – they have a degree but wonder if going further in their studies will help them in finding a better job. There is no clear answer to this. I was faced with this dilemma towards the end of my degree – many friends were doing Master's Degrees in Aquatic Resource Management in order to secure the best jobs, but I didn't go down this route; the costs were unrealistic, and I wanted to get to work! It is true that some technical and academic roles will ask for higher level education, so researching your preferred area of work is essential.

My career development was an example of a career 'curve' rather than change. I worked in Fisheries Management for UK Government agencies for six years and within that period, I changed the focus of my work from fieldwork and monitoring, to management, policy development and research projects. This came partly from promotions, but also from seeing opportunities for new projects and taking them. I also took professional examinations to boost my qualifications and knowledge base. My increasing involvement in policy and strategy development led to me moving to London to work in lobbying and policy development. The career jump from a fisheries role to a lobbying role could be hard to explain and achieve, but the fact that I had 'morphed' my past role to include policy development, lobbying and public liaison was a major benefit in my career move. My view was always to see how I might develop different areas of expertise in case another role came up in a different area. However, it was never just a calculating move; I really enjoyed broadening my skills and experience.

My career in policy development also involved finance, management and project roles, which further developed my knowledge. I became a bit of a 'jack of all trades, master of none', which is why my current role, specialising in consulting, careers and books, is a wonderful return to a greater focus on fewer work areas. The balance between 'generalism' and specialism is considered several times in this book, as it has often been a point of discussion with other environmental professionals.

There were several learning points from the 'middle' period of my career. I had a degree of movement in my profession, but I always stayed within the environment sector. The reason that I was able to do this was due to me taking on new areas of work whenever I could, without affecting my day-to-day job. In fact, many of the new skills I learned greatly benefitted my role and in addition helped me build multi-sectoral networks of new contacts. When a new job appeared, I already had a structured CV, which could be changed to match the numerous roles that were of interest to me. The 'next role' could have been financial, policy-focused, governmental or in project management – I had gained work experience and skills in all of these sectors through tasks that I had added to my current role. In addition, I learned an important skill - immerse yourself in the wider sector through networks, groups, newsletters and the like.

There is a significant debate between sector professionals, careers coaches and advisors over how much of a career plan you should have as well as the time span this should cover. My view, and my approach, is that a good practice to get into is to have a rough 'game plan', (not always strictly adhered to) which projects to the medium term, (perhaps five years) so that there is a focus for your career direction, network development, skills development and training. This can be a plan 'in mind', a structured mind map or document. Many careers advisors suggest a formalised career development plan, which aims to help in key areas: avoiding getting stuck in a rut; opening your eyes to new opportunities; having ideas of how to increase your skills and knowledge (such as training, taking on new projects and professional qualifications); and developing a clear sense of direction, which will give you confidence in your career progression. I have tended to avoid a formalised career plan because when I did have a developed one, it tended to give me a sense of irritation and impatience when I was behind the plan phases. However, for some people, a formal structure is very helpful.

Career planning is also a key moment to consider what you actually want to do. This is surprisingly rare as an action – many people are stuck in jobs that they do not really enjoy, from which they have no clear way out. If you project forward to your ideal role, you may be presented with areas of research and actions so that you can 'curve' your career in a positive way. This all sounds simple, but in practice, 'life' tends to get in the way; issues such as housing, family, finance and various other factors may make career change less possible. However, there are always ways in which you can continue to develop, ready for the next career step. Often, an ideal role falls right into your lap from nowhere and at other times, you hear of suitable roles from contacts or through sector research and networking.

I have been in the environment sector for more than 30 years and I feel that it is now time to pass on my experience to the next generation, which has a much more global outlook on career development. I mentor a group of young people in the environment sector, worldwide. A mentor is defined in the Cambridge English Dictionary (2021) as 'a person who gives a younger or less experienced person help and advice over a period of time, especially at work or school'. The practical examples of this are reviewing resumes and applications, suggesting suitable employers, course choices, interview help and sometimes direct approaches and recommendations to potential employers. There is also the element of personal support which is always encouraging. I would recommend to anyone in the sector that they take on this role. There are many courses on how to develop mentoring skills and many companies have their own training schemes in this area.

I was inspired to write about green careers after finding a book in the shop at the wonderful Monterey Aquarium in California, USA, whilst on holiday. It was 'Environmental Careers in the 21st Century' by The Environmental Careers Organization, published in 1999. It is a really helpful text covering green careers, mainly in the USA.

There is also a growing number of 'green career coaches' who can help with career development. The US expert, Carol McClelland, was always a great source of information in terms of personal development in the green sector.

1.3 Introduction to the Book

This book is aimed at anyone worldwide who has an interest in developing a career in the environment sector: students, undergraduates, career-changers, university and school careers services, major organisations that are asked for careers information and trade and professional body organisations in the environment sector.

There is a need for a resource that brings together a global text, as well as a linked online resource which cuts through the myriad jobs websites that one finds if you type 'environmental jobs' or 'green jobs' into a search engine, as well as the large number of books that examine a single country or a small part of the global environment sector.

This book focuses on the career paths that offer the greatest opportunities globally, specifically for those who want to develop an environmental career. There are, of course, many areas of work that are not covered. If I covered every sector, this book would be too big and unwieldy; areas such as mapping and GIS, wildlife crime and planning could easily have been included, but I aim to use the Env.Careers website to fill in any gaps and make any amendments needed.

This book is a practical guide to increasing your chances of developing a successful career within the global environment sector, wherever you are based and want to work. The book has a global multi-sector perspective and gathers information, tips and advice from professionals in the sector. The layout of the book uses as consistent a format as feasible and is written with a view to minimising jargon or acronyms. There are many differences in environmental terminology worldwide, which is very challenging, and I have tried to be open to using multiple wordings, but there are terminology differences that are unavoidable.

This book is supported by online content on the Env.Careers website, www.env.careers (managed by the author). As new sectors develop, the website will be the location for updates and background information. As the sector is maturing rapidly and changing swiftly, this book aims to meet the need for an up-to-date resource in this area, with online resources to back it up and provide updates.

This is an introductory book that is structured as a compendium of sectoral information. There are two larger chapters for water management and environmental consultancies, as these are key areas for people to develop environmental careers in globally.

There are frequent references throughout the book to informative website addresses. The author and contributors have attempted to ensure that, at the time of printing, these addresses were valid. However, websites are often dynamic, and addresses do change. With time, some of the addresses may no longer be valid. Nevertheless, I hope that, where this occurs, the references will still enable you to find your way to the same information, even if at a different address.

Please also note that a mention of a company, website or service within the book does not constitute an endorsement from the author or book contributors unless this is made very clear in the text.

1.4   What Is a Green Job?

There are many formalised as well as general definitions of what jobs fit into the global 'green' sector. Many are based on the use of the word 'environment', meaning 'environmental services', such as waste management, the supply of green goods and products like pollution treatment technologies. 'The Environment' covers the natural, built and human environment. There are many careers that touch on environmental issues but are not completely within the scope of being environmental. The type of positions and workplaces vary massively, as do the organisations working in the sector. Some of the many definitions are below.

OECD (2010) notes in its paper 'Green Jobs and Skills':

> For the purposes of this paper, green jobs are defined as jobs that contribute to protecting the environment and reducing the harmful effects human activity has on it (mitigation), or to helping to better cope with current climate change conditions (adaptation).

UNEP (2008) offers a more detailed definition:

> Green jobs are defined as work in agricultural, manufacturing, research and development (R&D), administrative, and service activities that contribute substantially to preserving or restoring environmental quality. Specifically, but not exclusively, this includes jobs that help to protect ecosystems and biodiversity; reduce energy, materials, and water consumption through high efficiency strategies; de-carbonise the economy; and minimise or altogether avoid generation of all forms of waste and pollution.

The Apollo-Alliance (2008) definition for so called 'green collar jobs' was

> Green-collar jobs are well-paid career track jobs that contribute directly to preserving or enhancing environmental quality. Like traditional blue-collar jobs, green-collar jobs range from low-skill, entry-level positions to high-skill, higher-paid jobs, and include opportunities for advancement in both skills and wages.

The US White House Task Force on the Middle Class (2009) noted:

> Green jobs involve some tasks associated with improving the environment, including reducing carbon emissions and creating and/or using energy more efficiently; they provide a sustainable family wage, health and retirement benefits, and decent working conditions; and they should be available to diverse workers from across the spectrum of race, gender and ethnicity (United States Department of Labour 2009)

The above definitions give a generally helpful outline as to what green jobs cover, but there are disputes over some roles. Often, there is a 'variety of shades' in terms of whether a job is green or not.

Brian Handwerk highlighted (2012):

> . . .defining exactly what green jobs are, how they can be created, and how they benefit the economy and environment presents quite a challenge.

As green issues become more integrated within company procedures and environmental regulation, there is an argument that trying to pigeonhole green jobs is not helpful. I can understand this argument – we want all jobs to have an element of being 'green' – but generally, green jobs are an obvious choice for those who want to actively 'make a difference' to the world we live in. There is a fundamental difference between someone who purely wants to just get a job and someone who wants to make a positive change in the world through their career.

In terms of government reporting, it is helpful to know which industry sectors are growing and where the key opportunities are. Often, government funding for green jobs is diverted into industries where the shade of green is questionable. However, Governments often struggle to decide what green jobs are and what should or should not be included.

Scholars at the Heritage Foundation took issue with the notion that government subsidies and spending are helping to create large numbers of new green jobs, according to a report by David Kreutzer (2012). He explained:

> The largest green jobs providers in manufacturing are steel mills (43,658 jobs) . . .Over 50 percent of all steel mill jobs are green. This high fraction of greenness is driven by the industry's reliance on scrap steel for the majority of its inputs, not by the greenness of the goods produced with the steel. The trend toward greater use of scrap steel is decades-long and is not the result of any green jobs initiatives.

Kreutzer also questioned the extent of green jobs and Government green subsidies in renewable energy:

> The electric power generation industry has 44,152 green jobs. . .This may seem like a lot, but only 4,700 are in renewable power generation, including 2,200 in wind, 1,100 in biomass, 600 in geothermal, and only 400 in solar. Though these totals do not include jobs in the manufacture or installation of these power sources, they pale to the equivalent green jobs count in nuclear (35,755), which accounts for over 80 percent of all green jobs in electric power generation.

The US Bureau of Labor Statistics' numbers above were reasonably close to the findings of a similar US study, 'Sizing the Clean Economy: A National and Regional Green Jobs Assessment', by the Brookings Institution (2011). However, the report does state:

> The clean economy remains an enigma: hard to assess. Not only do 'green' or 'clean' activities and jobs related to environmental aims pervade all sectors of the U.S. economy; they also remain tricky to define and isolate—and count.

The International Labour Organisation (2016), an agency of the United Nations, has published its own definition of green jobs in its article, 'What is a green job?' Their simple definition is

> Green jobs are decent jobs that contribute to preserve or restore the environment, be they in traditional sectors such as manufacturing and construction, or in new, emerging green sectors such as renewable energy and energy efficiency.
> Green jobs help:

- Improve energy and raw materials efficiency
- Limit greenhouse gas emissions
- Minimise waste and pollution
- Protect and restore ecosystems
- Support adaptation to the effects of climate change

In its Green Jobs Programme outputs, Gloria Acuña Navarro, from the National Learning Institute of Costa Rica noted:

> Green jobs can embrace environmental practices and decent work conditions to varying degrees. Enterprises can start with small/'lighter' green initiatives and with time move towards more comprehensive/'darker' green initiatives. Improving the environmental and social qualities of jobs is a continuous process that seeks to achieve sustainable development at national level.

Joanne Martens has developed the 'Green World of Work' Principle:

Figure 1.1 The Green World of Work. *Source:* Martens 2020. © Joanne Martens.

The Green World of Work (Martens 2020)

Change has impacted every aspect of our lives. There are tremendous forces reshaping society and with it, the world of work. Organisations and individuals are being required to accept their social, economic and environmental responsibilities. Sustainability and Resilience enable us to survive and thrive.

Environmental Sustainability

We face a number of environmental sustainability challenges in our communities, country and planet. Challenges are areas of focus that must be addressed to be sure that we have a healthy planet and to ensure that current resources will be in good shape for future generations.

Which Environmental Challenge Do You Care Most About?

Clean Water – Manage, deliver and treat water needed for industrial, agricultural, personal use and more responsible global use.

Communities – Build sustainable communities – neighbourhoods and cities – geared to human needs, so people can live and work in healthier ways.

Energy Efficiency – Increase energy conservation by reducing energy waste in our homes, businesses, cars and communities.

Food – Produce local organic food to keep people and the planet 'healthy'. Encourage safer agricultural growth and develop cleaner processes in food production.

Green Construction – Build homes and businesses that are energy efficient, are healthy to live in and minimise waste.

Green Transportation – Design, build and use alternative transportation systems that move people and goods in energy efficient ways. Expand virtual capacity to minimise travel.

Healthy Environment – Protect and rebuild our environmental resources – oceans, forests and water – and preserve the diversity of plants and animals.

Healthy People – Keep people healthy as environmental changes endanger human health worldwide.

Reducing Waste – Eliminate waste and decrease the use of landfills. Refuse one-time use products. Reduce consumption. Reuse materials and increase creative strategies. Recycle materials saves money and energy.

Renewable Energy – Generate renewable energy by harnessing wind, solar, geothermal and water resources.

Career Self-Reliance

Career Self-Reliance is a lifelong commitment to actively manage your work life and learning in a rapidly changing environment. 'Work is what you do to learn a living' (Career Action Center 1997).

How Are You Developing the Characteristics of Career Self-Reliance?

- *Self-Aware* – You know who you are, where and how you do your best work; you understand and can articulate the value you add.
- *Values-Driven* – You have determined the values that give direction and meaning to your work.
- *Dedicated to Continuous Learning* – You regularly benchmark your current skills and create a development plan to keep your skills current.
- *Future-Focused* – You look ahead to assess customer needs and business trends; you consider the impact of those trends on your work and on your development plan.
- *Connected* – You maintain a network of contacts for learning and sharing ideas; you work collaboratively with others towards mutual goals.
- *Flexible* – You anticipate change and are ready to adapt quickly.

World of Work

The concept of work is employment.

- The structure of work are companies, organisations, governments and communities.
- The work that needs to be done is often defined by jobs, roles, occupations, etc.
- The workforce is made up of people who possess the knowledge, skills and attitudes to perform the work.
- The work environment is physical, virtual and cultural.

How Do You Fit into the New World of Work?

The New World of Work redefined the social contract of long-term employment, a dependent relationship between employer and employee to a social agreement between organisations and workers, developing interdependent relationships and fostering a resilient workforce. Instead of one career for a lifetime, the norm is between three and five careers in a lifetime.

Changemakers

Changemakers of the green generation are playing important roles in all challenge areas to achieve environmental sustainability. Changemakers work in government, business and industry, non-profit organisations and research and education institutions.

What Changemaker Role Most Interests You?
- *Policymaker* – Guide people to make decisions and take action to do the right thing.
- *Innovator* – Work with ideas, research facts and develop new methods, products or designs.
- *Communicator* – Educate and inform others about issues, solutions and products.
- *Implementer* – Get the job done and make solutions a reality.

Fundamentally, if your career path is to pursue a green career, you may have a reasonable idea as to the type of jobs you would like to seek. However, the confusion over definitions means that the targeting of 'green funding' by governments can often miss the target in terms of driving green career development and the number of truly 'green' jobs.

1.5 The Global Green Jobs Sector

The environment is a growing sector for employment globally, and there is also a growing public awareness of green issues, on a global scale. The diversity of green jobs has accelerated, with new sectors of employment being stimulated by technological advances (often called 'Green Tech').

Green jobs span a wide array of occupational profiles, skills and educational background and, while some constitute entirely new types of jobs, the majority are modified versions of traditional professions and occupations. There is clear evidence of the viability and potential for green jobs across the entire workforce, including manual labourers, skilled workers, craftsmen, highly qualified technicians, engineers and managers.

Creating and Stimulating Global Green Jobs Growth

There has been considerable inertia and a global resistance to change in terms of developing global green jobs. The UN has been pivotal in terms of highlighting this issue and stimulating funding for international programmes to encourage green jobs growth.

There are significant barriers to growth in green employment programmes. Some areas have poor infrastructure, limited policy measures and little political will to actively develop green sectors. Climate change is having a direct impact on jobs and livelihoods; some sectors, which are seen as being energy inefficient and damaging environmentally, have major employment, so this is a challenge that is yet to be faced. There needs to be a global transition that garners support from all sectors and governments.

The UNEP report from 2008, 'Green Jobs: Towards decent work in a sustainable, low carbon world' noted:

> This report shows for the first time at global level that green jobs are being generated in some sectors and economies. This is in large part as a result of climate change and the need to meet emission reduction targets under the UN climate convention.
>
> This has led to changing patterns of investment flows into areas from renewable energy generation up to energy efficiency projects at the household and industrial level. The bulk of documented growth in Green Jobs has so far occurred mostly in developed countries, and some rapidly developing countries like Brazil and China.

The US Bureau of Labor Statistics' data from 2009 did not attempt to track how many green jobs were new and had been created by newer 'green jobs' programmes, but the scheme to measure green jobs was shelved in 2013 due to budget cuts. It did report that 2.3 million jobs were in the private sector, while 860 000 were public jobs (US BLS 2013).

Reuters highlighted in 2016: 'China expects the output value of its energy saving and environmental protection industry to rise from 4.5 trillion yuan ($653 billion) last year to more than 10 trillion by 2020, lifting its share of gross domestic product from 2.1 percent to 3 percent... the industry currently employs more than 30 million people... China is hoping that big outlays in the environmental sector will not only reverse some of the damage done by more than three decades of breakneck economic growth, but will also help diversify the country's heavy industrial economy.'

In China, as well as in other countries, there is an impetus for growth in global green jobs, through managing issues like climate change and poor air quality, transitioning heavy industries and also developing new opportunities for global trade.

The South China Morning Post reported (2018) that 'China is looking to set tougher goals in a new three-year "green" plan to improve air quality and tighten regulations' and this has not only restricted certain industries but has also stimulated investment in environmental technologies.

Mark Muro, from the Brookings Institution report (2012), reported that:

> The 'green' or 'clean' economy exists; that we can define it; that in fact we can count its jobs and measure it and track its progress. Given that, it is soon going to be time for election-year combatants to leave aside both flat-earth denials and exaggerated claims and get down to realistic dialogue about how to foster what turns out to be a modest-sized, manufacturing-oriented, unavoidable piece of America's next economy.

His team's study reported that most green jobs in the United States are in already-mature segments, like manufacturing or public services, including wastewater and mass transit. A smaller proportion includes newer green energy ventures such as solar, wind and smart grids – but in recent years, these sectors have been adding jobs at rates that far outstrip national averages.

The 2008 report 'Working Towards Sustainable Development', by the Green Jobs Initiative (United Nations Environment Programme (2008), the International Labor Organization, and the International Trade Union Confederation), noted:

> Most studies indicate gains in the order of 0.5–2 per cent, which would translate into 15–60 million additional jobs globally. More ambitious green growth strategies could result in even stronger net gains in employment by triggering a wave of new investment into the real economy, as suggested by studies of Australia and Germany'. . .A significant potential also exists in emerging and developing countries. For example, targeted international investment of US$30 billion per year into reduced deforestation and degradation of forests (REDD+) could sustain up to 8 million additional full-time workers in developing countries.

The International Labour Organisation (ILO) (2018) has formed a Green Jobs Global Team, and in their 2018 report 'Greening with Jobs – World Employment and Social Outlook 2018', they state:

> 18 million jobs can be created by achieving sustainability in the energy sector. Limiting global warming to 2°C by the end of the century will create, by 2030, jobs in construction, electrical machinery manufacturing, copper mining, renewable energy production and biomass crop cultivation. But there will be fewer jobs in petroleum extraction and refinery, coal mining and production of electricity from coal meriting policies to protect these workers to make sure the transition is just for all.

Santander (n.d.), the multinational financial services company, outlined in its release 'The future of employment is green':

> If only a couple of decades ago, in the early years of the new century, we would have asked someone about what green employment means, it is most likely that the response obtained would not have implied much more than a park ranger or something similar. And it would not be a bad option, but it would remain insufficient.
>
> That same respondent would have thought we were crazy if, predicting the future, we would have advanced to the following figures: 400,000 new jobs in Europe, more than 50,000 only in Spain and almost 18 million around the world. That is the estimate made by the European Commission in the case that all the legislation enforced about waste would be applied in a real way.
>
> Today, green employment is becoming, with every passing second, a magnificent job opportunity. So, roughly speaking, we could define it as that which has as an absolute objective so lofty as protecting the planet. And how can such a laudable goal be achieved? Basically, by reducing the impact that human action itself, especially through large companies, provokes in the natural environment.

Within this sector, the most demanded areas are those related to energy efficiency, waste management, pollution prevention and many others included in what we know as Corporate Social Responsibility of companies. Ecological agriculture, environmental communication, sustainable tourism and eco-design are some of the most innovative fields that green employment already places in a powerful place of the supply and demand market.

Box 1.1   Diversity in the Environment Sector

There is an urgent challenge in the sector that in most developed nations, there is significant under-representation in the environment sector of minority groups and women in terms of engagement with nature as well as in employment.

A headline from the UK body Natural England (2019) in its national survey 'Monitor of Engagement with the Natural Environment' noted:

> Children from black, Asian and other minority ethnic backgrounds are less likely to spend time outdoors than children from white backgrounds...57% of children with black, Asian and minority ethnic family backgrounds spend time outdoors at least once a week, compared with 73% of children from white family backgrounds The majority of children (72%) had visited urban greenspaces in the last month while just over a third had visited the countryside (36%) and 17% visited a coastal location. The results for young people show a similar pattern: the majority of visits taken by 16- to 24-year-olds were to urban greenspaces (62% of visits), with smaller proportions to the countryside (28%) or coast (11%).

The think tank Policy Exchange (2017) also researched diversity in different occupations. They found in their research that, of the 202 occupations, farmers were the least diverse and environment professionals were ranked 201.

In a key 2014 US report by Dorceta Taylor (2015), commissioned by Green 2.0, 'The State of Diversity in Environmental Organizations' she highlights:

> The current state of racial diversity in environmental organizations is troubling, and lags far behind gender diversity...Environmental jobs are still being advertised and environmental organizations recruit new employees in ways that introduce unconscious biases and facilitate the replication of the current workforce...Moreover, environmental organizations do not use the internship pipeline effectively to find ethnic minority workers

This challenge is being met by many organisations and individuals who campaign and have activities in the area. These include

Green 2.0
www.diversegreen.org

(Continued)

Box 1.1   (Continued)

Black2Nature
www.yearofgreenaction.org/green-actions/black2nature

Bird Girl
www.birdgirluk.com

Diversity Joint Venture for Conservation Careers (DJV)
www.diversityinconservationjobs.org/about

Center for Diversity & the Environment
www.cdeinspires.org

Student Conservation Association - Career Discovery Internship Program
www.thesca.org/cdip

SACNAS – Society for Advancement of Chicanos/Hispanics and Native Americans
in Science
www.sacnas.org

Women for Wildlife
www.womenforwildlife.com

Changing Patterns in 'Hot Topics' and Funding

If your aim is to develop a long-term career in the environment sector, there are significant challenges in the fact that the areas of work that attract funding, and therefore offer more jobs, shift from year to year. Government funding priorities can have an immediate impact on the sector, such as announcements on support for areas such as wind power and solar. The classic way to respond to this is to become a generalist, with skills that enable you to shift across areas, but ultimately if your aim is to be, for example, a fisheries or ecology professional you need to develop your skills to enable you to develop a career in the area and to ride out any periods of recession in your sector. In the early 1990s when I graduated, there were few jobs being created in fisheries, but there was still 'movement' in the sector which enabled me to get a job and get on the career ladder.

The media hot topic of plastic pollution has increased the profile of water pollution, life cycle analysis, materials production, waste minimisation and management and recycling, but it has had minimal effect on global green jobs. However, the need for solar and wind energy, for multiple reasons including energy production diversity, lower energy costs, climate change response and non-renewable resource limitations, has led to a rapid increase in global employment. The drivers for green jobs growth are many and varied. As noted earlier, a sensible approach is to develop skill sets which enable you to respond to sectoral changes but stick to the main core of the roles which led you to work in the green sector in the beginning.

However, there are a number of sectors displaying increasing investment and global jobs growth which are worth exploring if you want to develop a green career.

Fastest Growing Sectors

Renewable Energy
The renewable energy sector has experienced significant global growth in recent decades. As noted by Deloitte in their 2018 Deloitte Insight 'Global Renewable Energy Trends' article, renewables are becoming a 'preferred' energy choice globally. Deloitte's report notes seven 'enabling trends' and 'demand trends' driving this process:

Enabling trends: price and performance parity, balancing the grid and new technology and innovation

Demand trends: smart cities, community energy, new emerging markets and growing corporate energy buying

Renewable energy offers a very good opportunity for jobs to migrate across to the green sector from traditional heavy industry roles. The sector has roles in areas, including manufacturing, site installation and engineering design, as well as research, environmental management, and impact assessment. This 'role transferability' makes it an attractive choice for governments globally, as the transition process for jobs, skills and infrastructure is streamlined, compared to implementing very 'different' new technologies. This book has a chapter on Renewables and Energy.

Fastest Growing Sectors

Environmental Consultancies
There has been a rapid growth in consultancies, especially global environmental and engineering consultancy firms, who have diversified their areas of work. Many have grown significantly through mergers and acquisitions. These mergers continue as key players in the sector seek to secure their positions as 'super consultancies'.

The so-called 'Global 23' companies (now called the 'Global 22' by Environment Analyst in 2019) dominate the global market with over 40% of the total global market. These companies include AECOM, Tetra Tech and Arcadis.

The services delivered by consultancies have developed significantly. There are a small number of global consultancies who are able to manage virtually any engineering, technical and environmental service needed, from building a dam or bridge to developing policies and strategies for governments. There are also many more specialised consultancies who offer services in more specific areas such as ecological surveys, planning, policy development and environmental restoration works. Increasingly, the more successful of these smaller consultancies are being swallowed through acquisition by larger consultancies who want to be able to offer further services and gather the clients of the smaller companies. There is also a pool of experienced independent individual consultants who are hired as freelancers by the larger consultancies as and when they are needed.

Consultancies have become increasingly global in their reach, operations and main offices. This also rings true for employment within these companies – many multinational organisations have teams of experts who work on a global basis, as well as considerable numbers of local staff. This opens an opportunity for global environmental careers in a diverse range of areas of employment. However, the limitations of working rules and visas for certain nationalities can complicate the flexibility for some staff to secure global roles within consultancies. This book has a chapter on Environmental Consultancy as this is a key area to consider for a green career.

1.6   The Limitations of Online Searches

The Internet is an amazing resource for job seekers, especially those in such a diverse field as the environment. The process of job hunting and linking to employment agents and jobs websites has become streamlined and simplified. Unfortunately, as many of us are beginning to realise, using Internet searches for many elements of green career development is becoming an exhausting process unless you're willing to trudge through what could be termed 'dross' all day, seeking a single gem of a site. It is particularly complicated for those who have not yet focused in on a specific area of work.

Information on global environmental careers, both in print and online, is messily scattered and most is very out of date. Most information is either UK-centric or aimed at US career hunters. There are few resources available that outline in an organised and selective manner what sectors are available and how to get a job in those sectors. A very high proportion of roles and internships in some sectors are not advertised externally and some are only visible on specialised blogs or mailing lists.

Many online resources are just jobs boards, with no supporting information for a career hunter. Where there is careers information, it is often out of date and limited in scope. Even the most promising online resources are hard to find and are not securely funded, often depending on short term charitable funding.

In the current maturing environment sector, there are clearer routes to environmental jobs, and careers services have more computing and search power in their hands to advise on the sector, although careers services worldwide tend to be under appreciated and under resourced. In addition, the internet has helped with job and information finding, but it has also caused a decline in information quality and a clogged internet of old ideas and urban myths.

National governments have supported numerous jobs websites in order to encourage growth in the sector. In the United States, the Government-supported website Careeronestop (www.careeronestop.org) aims to provide comprehensive careers information in a similar way to the National Careers Service in England (www.nationalcareers.service.gov.uk). However, because these websites are a 'catch all', finding the information required and then jobs available in the sector can become an exhausting process.

The global environment sector is growing rapidly and there are millions of roles which are open to those who can use the right resources to get them into the sector. Competition for environment sector jobs is very fierce and job hunters need the right advice, resilience and 'stickability'. Rapid change in the sector leaves career professionals playing catch up in terms of being supportive to job seekers.

1.7 So Where Do I Start?

There is a blinding amount of information available, some of which will be helpful and much which will take you in circles. There are some fundamental tasks which you should undertake. The tasks below are not in a specific order and some are less relevant to those at different stages of their career.

Tasks for those before 'working age':

1) Do research into the sector. Do you have personal interests that make you want to develop a green career, such as nature watching, fishing or conservation volunteering? Look at roles within the sectors that are relevant to your interests and see what skills, education and knowledge they need. Sign up to relevant newsletters and news services.
2) Consider voluntary work and look at organising relevant work experience – this can help with applications for further education as well as looking good on a CV or resume. It also can assist you in developing skills and knowledge, which are helpful in any career, let alone a green career.
3) Consider what school and college choices give you options to develop a green career.
4) Use social media as a tool for information gathering, as well as a way to develop networks of contacts. Getting into the habit of developing and maintaining a network is sensible.

Tasks for those of 'working age':

1) Sign up to jobs email and update services and link up with jobs agencies. Even if you are new to the sector or have minimal qualifications and experience, it is worth getting to know the sector deeply. If you have a specific area of interest, 'immerse' yourself in the subject through newsletters and other news services.
2) Review your current situation and organise your CV or resume. Many roles ask for specific knowledge and qualifications, so this will help you target your search and consider future training and education. There are many books and websites that focus on CV and resume styles. There are also many experts that may be willing to help you.
3) Consider using social media and professional networks that are helpful. On LinkedIn, there are relevant networks which include Environmental Careers Network, Sustainability Career Group, Green, Environmental Consulting Professionals and Green Jobs & Career Network.
4) Is there a mentor who can help you? Mentoring is growing in scope and not just for people already within a company mentoring scheme. LinkedIn have a scheme for mentoring.

Tasks for those who want to transition into a green career:

1) Transitioning is not necessarily a major career change for many, merely a migration into the area using current skill sets. Use online tools to see what roles you can move into and

review your skills and CV or resume to consider what elements would be helpful for a green career. Look at job adverts for roles that would interest you and develop your skills to match their requirements.

2) Consider additional training to help you fit into the right career.

Each section of this book has information on training and helpful resources in that sector.

1.8 Volunteering

Many careers in the sector started through volunteering for organisations such as nature trusts, conservation NGOs (Non-Government Organisations) and other small bodies. Volunteering is an option for graduates to build up skills and knowledge, as well as for those who are pre-University. There are also many organisations specifically focusing on volunteering. The table below lists many organisations where you can volunteer, and in addition some offer summer jobs, placements, internships and full-time employment. Some volunteering is organised commercially, so this may not be free of cost. There have been concerns raised at the growth of paid-for volunteering and unpaid internships.

Sarah Bell (2015) wrote her Master's Degree thesis on volunteer tourism in the 'conservation' industry, and her paper gives a helpful insight into the motivations for conservation volunteering.

The Guardian (2017) highlighted concerns over paid and free volunteering and noted:

> a depressingly similar story: serial unpaid internships, crippling student debt, short-term work for little or no pay, dismissive attitudes, and entry-level job requirements that include expectations of considerable field time and experience. . .Exacerbating the dismal job market is this trend of graduates becoming stuck in full-time unpaid internships or long-term volunteering.

The hope is that volunteering and internships will be a positive rather than exploitative experience.

Box 1.2 Volunteering Opportunities in the Environment
Adventures in Preservation
Different 'adventures' for volunteering projects for students to get involved in, in different locations across Europe. Support provided too through staff to endure safety and enjoyment of those on placements www.adventuresinpreservation.org
BC Parks, Canada www.bcparks.ca/volunteers
Backdoorjobs.com Short-term job adventures, many in environmental areas of work www.backdoorjobs.com

Blue Ventures
Marine conservation organisation situated within Madagascar, enabling people to volunteer to protect marine coastline and work with the local community
www.blueventures.org/about

British Council
Placements abroad for UK residents
www.britishcouncil.org/study-work-abroad/outside-uk/iaeste

BUNAC
Opportunity to work/volunteer/intern abroad – a particular focus towards Africa for conservation roles. Placements available over different timescales (chance to go over summer in-between studies as well as after graduation)
www.bunac.org/uk/volunteer-abroad/wildlife-and-conservation

Conservation Jobs
www.conservationjobs.co.uk/volunteers/

Conservation Volunteers New Zealand
www.conservationvolunteers.co.nz

Earthwatch
Conservation projects
www.earthwatch.org/europe

Environmentjob.co.uk Volunteering listing
www.environmentjob.co.uk/volunteering/

European Outdoor Conservation Association
European Association for volunteering projects and ways to get involved (links to organisations and roles students/young people can apply to). Able to go for short-term or long-term (employment/internships)
www.outdoorconservation.eu/conservation-detail.cfm?pageid=14

Friends of the Earth
More centred towards campaigning rather than physical volunteering, but still looks appealing to those unable to travel far but still wish to make an impact on conservation www.foe.co.uk/page/job-vacancies

Frontier
www.frontier.ac.uk/

Green Volunteers
Conservation projects
www.greenvolunteers.com

Greenforce
Conservation projects
www.greenforce.org

(Continued)

Box 1.2 (Continued)

Greenpeace

Global NGO with wide variety of issues on environment (with ultimate focus on climate change) seen as 'radical' in comparison to other NGO's but very interested in volunteering/hiring students

www.greenpeace.org.uk/what-you-can-do/jobs

IFAW

Organisation aimed to help save individual and populations of animals at risk of extinction. Website provides information on projects students can get involved with.

www.ifaw.org/united-states/about-ifaw/employment

Kaya Responsible Travel

Volunteering opportunities across Asia, Africa and the Americas. Able to undertake various roles, including marine and wildlife conservation

www.kayavolunteer.com/

National Park Service

Volunteers-in-Parks programme

www.nps.gov/getinvolved/volunteer.htm

National Trust

www.nationaltrust.org.uk/find-an-opportunity

Natural England

Conservation within England (mainly centred around habitat restoration) but with strong links to conservation schemes across the country

www.gov.uk/guidance/volunteering-with-natural-england-how-to-get-involved

Operation Wallacea

International organisation (only available to students at university)

www.opwall.com/

Ornithology Exchange

Seasonal positions, internships, volunteer opportunities and more

www.ornithologyexchange.org/jobs/board/short-term-positions/

Peace Corps

Primary focus towards human aspect but various opportunities for environmental conservation available

www.peacecorps.gov/volunteer/learn/whatvol/env/

Raleigh International

www.raleighinternational.org/volunteer/

Royal Geographical Society

Information and advertisements on graduate roles in conservation volunteering opportunities.

www.rgs.org/OurWork/Study+Geography/Careers/Finding+jobs.htm

RSPB
British Association towards bird protection – various volunteering conservation roles available across Britain
www.rspb.org.uk/vacancies/

Steph into Nature
www.stephintonature.com/blog/volunteer-for-wildlife-online

The Conservation Volunteers (TCV)
Restoration activities across the UK to protect green spaces. Positions available over short-term so beneficial to those looking for opportunities whilst studying
www.tcv.org.uk/

The Dirty Weekenders, Scotland
www.dirties.org.uk/about/

The Great Projects
Focus towards animal conservation primarily across Europe but also options more global. Website offers key information about flights and insurance needed
www.thegreatprojects.com/volunteering-projects

The Nature Conservancy
www.nature.org/en-us/get-involved/how-to-help/

The Royal Parks
www.royalparks.org.uk/get-involved/volunteer-with-us

The Student Conservation Association (SCA)
General information for students interested in conservation. Youth programmes including volunteering
www.thesca.org/about

The Wildlife Trusts
www.wildlifetrusts.org/closer-to-nature/volunteer

UN Volunteers
The United Nations Volunteers (UNV) programme contributes to peace and development through volunteerism worldwide
www.unv.org/

Volunteering Matters
Action Earth and other programmes
www.volunteeringmatters.org.uk/project/action-earth/

VSO Volunteering
www.vsointernational.org/volunteering

Wilderness Volunteers
www.wildernessvolunteers.org/

(Continued)

Box 1.2   (Continued)

Wildlife Conservation Society

Aim of increasing global biodiversity (particularly with charismatic species). Website lists targets for preserving each listed species and current issues surrounding conservation
www.wcs.org/about-us/careers

Woodland Trust

UK-based charity – interested in those looking for volunteering roles, as well as opportunities in graduate schemes/careers
www.woodlandtrust.org.uk/get-involved/

Working Abroad

Organisation allowing users to search through conservation volunteering schemes across the world
www.workingabroad.com/volunteer-organisations/

World Wildlife Fund

Internationally recognised organisation in protecting species (as well as emphasis on habitat restoration/preservation)
www.wwf.org.uk/about_wwf/jobs/

Worldwide Workers on Organic Farms
www.wwoof.org.uk

To note: if the above organisations do not have suitable volunteering available, they are also key employers in the sector, so are worth keeping an eye on, nonetheless.

1.9   Green Careers Coaches and Green Career Services

Green Career Coaches are a relatively new and growing phenomenon within the well-established area of career coaching. Career coaching was defined by Donna Sweidan, in a 2013 interview in Forbes (2013) magazine:

> I approach it as a discipline composed of two similar but distinct tracks: coaching and counseling. The goal is to support people in making informed decisions about their career development and trajectory, as well as offer various tools that they can use – résumés, cover letters, LinkedIn profiles – to meet those goals. . .definitions of the field – and the work – may still vary among more conventionally trained coaches.
>
> In general, 'coaching' tends to be a solution-oriented approach, which involves working with clients to see what concrete steps they can take to achieve career objectives. 'Counseling', however, is more process driven – you look at whether there are any behavioral, emotional or psychological issues that could be impeding a person's desired career ambitions.
>
> But the core virtue of career coaching is to help people assess their professional situations with a greater degree of honesty, curiosity, empathy and compassion.

Green coaches can help you to develop a career in the sector, but they differ in many ways, including

Approach

Some focus on your approach to life and your personality in order to help you develop a fruitful career. Others have a focus on directing you towards a green career through professional advice, training and linkages to main employers.

Clients

Their clients vary between graduates, career changers, senior executives and all levels of management.

Sectors Covered

Can be highly focused such as conservation only, or can cover corporate sustainability, or the entire green careers sector.

Location

Normally country-based – many focussing on the UK or USA but can be online and/or global.

1.10 Background

Backgrounds vary for career coaches. Many have university careers services backgrounds or professional experience in the environment sector. Some are psychologists or councillors.

Services Offered

Services from coaches can include:

- Packages of training
- Careers advice
- Subscriptions to ongoing newsletters
- One-on-one coaching interviews
- Events and webinars
- Helping with professional network listings, resume development and cover letter advice
- Interview skills
- Salary negotiation
- Offering professional link ups (connecting people)
- Job search skills
- Careers advice for universities and developing internship programmes for companies as well as being corporate recruiters
- Some coaches offer a bespoke service that includes adventure-based trips and inspirational travel.

Coaching Qualifications and Certifications

There are many different types of career coaching certifications. The main thing is to ensure that the career coach has a certification which requires them to adhere to a strict code of professional conduct for the career coaching profession. However, some companies offer a career advisor service from a professional standpoint, so they are more likely to have professional sector qualifications rather than coaching qualifications.

If additional help is required beyond career coaching (for example, mental health counselling), it is important to seek a Licensed Professional Counsellor (or equivalent) which requires a master's degree (e.g. in Careers Counselling) with additional training.

Listed below are a number of the career coach, career counsellor and coach instructor credentials (but there are many others):

Center for Credentialing & Education (CCE)
www.cce-global.org/credentialing

CCE – Global Career Development Facilitator
www.cce-global.org/gcdf

National Association of Colleges and Employers
www.naceweb.org/professional-development/coaching-certification/

National Career Development Association (NCDA)
www.ncda.org/aws/NCDA/pt/sp/credentials

Association for Coaching
www.associationforcoaching.com/

Institute of Leadership and Management
www.i-l-m.com/learning-and-development/coaching-and-mentoring-qualifications

Career Development Institute
www.thecdi.net/Getting-Qualified/Masters-and-other-courses
www.thecdi.net/Qualification-in-Career-Development

Should I Consider Using a Green Career Coach?

A key issue is that career coaches are professionals and as such they charge fees, depending on the service given. This will preclude some people from considering using a coach to help them.

Career coaches have a wealth of information and experience and many have detailed insights into the organisations you may want to develop a career with. As in many issues, one needs to research the area and then decide based on affordability and need.

The table below lists some of the coaches available.

Box 1.3   Green Career Coaches and Advisors

Careerfolk LLC
www.careerfolk.com
Donna Sweidan

Careers for Social Impact
www.careersforimpact.com

Catriona Horey
www.withcatriona.com
www.linkedin.com/in/catrionahorey

Charly Cox
Founder of The Climate Change Coaches
www.charlycox.com

Climate Change Coaches
www.climatechangecoaches.com/our-team

Conservation Careers
www.conservation-careers.com
Nick Askew

Environmental Career Coach
www.theenvironmentalcareercoach.com
Laura Thorne

EnvironmentalCareer.com
www.environmentalcareer.com
www.environmentalcareer.com/coach
John Esson

Green Career Advisor
www.greencareeradvisor.com
Lisa (Yee) Yee-Litzenberg

Green Career Tracks
www.greencareertracks.com
Barbara Parks

Green Growth Coaching
www.greengrowthcoaching.com
Charlotte Lin

GreenMatterZA
www.greenmatterza.com
Janavi Da Silva
www.linkedin.com/in/janavi-da-silva-pr-sci-nat-a90409aa

Hamish Mackay-Lewis
www.hamishmackaylewis.com

The Intelleto Advantage
Joanne Martens
joanne@theintelletoadvantage.com

(Continued)

Box 1.3   (Continued)

Green 360 Career Catalyst
www.green360careercatalyst.net

Megan Fraser
www.meganfraser.org

New Lens Consulting
www.newlensconsulting.com
Sharmila Singh

Nonprofit Career Coach
www.thenonprofitcareercoach.org
Mark McCurdy

Sustainable Career Pathways
www.sustainablecareerpathways.com
Trish Kenlon

Walk of Life Coaching
www.walkoflifecoaching.com
Shannon Houde

We Are All Wonder Women
www.weareallwonderwomen.com
Dr Eugenie Regan and Nadine Bowles Newark

Zoe Greenwood
linkedin.com/in/zoegreenwood

Top Tips and Profiles – Green Career Coaches and Services

◀ **Top Tips**

Lisa Yee-Litzenberg, President, Green Career Advisor, USA

1) Follow your heart. Focus on environmental issues you are passionate about.
2) Build relevant experience on these issue areas through internships, volunteer work, project work, etc. It's not enough to only have book smarts. Employers will not hire you unless you demonstrate real-world experience.
3) Never stop learning. Learn new skills that will make you valuable in your field, read what your competitors are doing well, keep up on new technological developments and career trends and join professional associations.
4) Build strong relationships with people in your field. People always remember how you made them feel.
5) Pay attention in each role to what skills you are good at and that you enjoy using. These are the ones you will want to seek out in future roles. It's always good to be thinking about your next career step so you can be building the right skills along the way.

Personal Profile

As a kid, I spent a lot of time playing outdoors and catching frogs and insects. This led me to develop a love of the natural world. In college, I was not sure what I wanted to do until I took a class on tropical rainforests and learned that they contained the greatest biodiversity on earth and that many rainforests were being destroyed. I decided right there in that class to dedicate my life and career to protecting endangered species.

My first job out of college was working for National Wildlife Federation giving endangered species presentations to colleges in 12 states and training students to take action to protect endangered species. I then worked on a project to protect Lake Superior. Later I served as a liaison between NWF and four state affiliates working to strengthen the state groups by strengthening the diversity of their finances and leadership. My last role at NWF was as the Program Manager for the Great Lakes Wolf program.

I then went on to lead the career services at University of Michigan's School for Environment and Sustainability for 10 years. In 2016, I launched my own business Green Career Advisor to help people find and secure their dream green jobs.

While leading the career services for environmental students at University of Michigan, I developed a list of what I called Universal Job Skills by reviewing hundreds of green job postings across all sectors and creating a list of the most sought-after skills by green employers:

- Strong communication skills: writing and public speaking
- Ability to work in multidisciplinary teams
- Strong interpersonal skills
- Management skills (budget management, project management, people management, change management, time management, fundraising, strategic planning/vision)
- Leadership skills/showing initiative
- Analytical/creative problem-solving skills
- Technical skills (software, field skills, graphic design, etc.)

Source: Lisa Yee-Litzenberg, President, Green Career Advisor, USA, www.greencareeradvisor.com. © John Wiley & Sons.

Organisational Profile

Climate Change Coaches

The Climate Change Coaches provide upbeat, compassionate coaching for organisations and individuals and run training programmes on how to use coaching skills to empower and influence others. They offer confidential conversations, free from judgement and unwanted advice, to help people think clearly about climate change and then get into action to do something about it.

Their team are all coaches and come from backgrounds in leadership development, psychotherapy and change management and have worked in a broad range of fields including in sustainability, environmentalism, international development, the military, start-ups and multinational for-profit businesses. In addition to offering workshops and personal coaching, they share a quarterly newsletter filled with examples of positive and inspiring action for the environment, alongside thought pieces about the psychology of change.

Source: Adapted from www.climatechangecoaches.com/contact-us
www.linkedin.com/company/climatechangecoaches
Twitter: @ClimateCoaches. © John Wiley & Sons.

Top Tips

Laura Thorne, aka The Environmental Career Coach, USA

1) Make sure your resume, cover letter and LinkedIn profile are all working and working together, to help you stand out and get your next interview.
2) Create a networking strategy that helps you to build momentum and builds your confidence. The more you converse with others on a consistent basis, the more comfortable you will become, and the more you will see how much value you bring to the conversation.
3) Do not settle, figure out what you want and make a plan to get it. Create an action plan, get someone to hold you accountable to it and review your progress on it routinely.
4) Look within yourself to discover what you want to do. Asking others what you should do is like asking them to choose your meal at a restaurant; you may not like what you get and you'll likely blame them for the bad choice.
5) Continue to map your career plan even after getting hired. People who don't do this often find themselves feeling stuck a few years down the road.

Personal Profile

Laura works with students, recent grads and career changers from all backgrounds who are seeking environmental careers to gain confidence in their searches by crafting sound career navigation plans. Having been a biology student and career changer and now a business owner, her coaching comes from experience, a passion for science and a commitment to the well-being of our planet.

She is well connected across the United States in the environmental and sustainability fields through her position as past president of the Tampa Bay Association of Environmental Professionals and now serving as a board member of the National Association of Environmental Professionals. These leadership positions and her experience as a grant manager for habitat restoration projects means that she has exposure and experience in a wide variety of both public and private industries.

Laura has a Bachelor's Degree in Biology from the University of South Florida, is a certified Project Management Professional (which means she is an expert in creating actionable plans) and has walked in your shoes!

Source: Laura Thorne, aka The Environmental Career Coach, USA, www.theenvironmentalcareercoach.com. © John Wiley & Sons.

Personal Profile

Joanne Martens, Owner, The Intelleto Advantage, USA

Joanne took the road less travelled. Navigating her career trajectory with five careers in a variety of industries, Joanne worked in corporations and start-ups, non-profit and academic institutions, as well as running her own business. Blending these experiences with her passion for imagining future possibilities and inspiring human potential Joanne honed her expertise of coaching and consulting in Leadership Development, Career Development, Workforce Development and Change Management.

Joanne is a recognised thought leader in pioneering career transition strategies and developing the concepts and practices of Career Self-Reliance and Workforce Resilience. She is passionately dedicated to lifelong learning and developing future leaders around the world. She coaches executives, leaders and teams and facilitates Women in Leadership Group Coaching Programs.

Forecasting the profound impact of climate change and experiencing transformation of the world of work, Joanne founded and co-developed Green 360 Career Catalyst to inspire, educate and empower young adults around the world to become contributing members of the green workforce.

Joanne is a student of life, the world her playground. Travel is her gateway to adventure: exploring the great outdoors and learning about global communities and cultures.

◀ **Top Tips**

Know Yourself – the Whole of the Who that is YOU!

BE...
- driven by Purpose
- guided by Vision
- ignited by Passion
- directed by Mission
- committed to Action
- fulfilled by Impact

Define Your Best Fit Job

Begin by defining your personal characteristics and criteria

What work do you want to do?
- Identify the skills, knowledge, abilities you have; the experience you want to leverage

Why do you like to work?
- Identify your interests, passions, motivations, values and purpose

How do you like to work?
- Identify your personality type and leadership style
- Identify working relationships
- Identify kind of work environments and organisational culture

Your specific selection in each of these categories combined become Your Ideal Job
  Preferences
Use them as
- your anchor, being grounded in who you are and the work you want to contribute
- your compass, guiding you with clarity and focus to where and with whom you
  want to work.
Do this, so you don't go looking for success in the wrong places, and you don't put
  yourself in places you don't belong.
Connect with others – Build Relationships
Connecting with others is an essential life skill and career strategy. It's about building
  relationships with people who guide and support you along your career journey.
You know who you are, time to get in the groove.
You have everything you need to make your next move.
Oh, the people you'll meet will take interest in you, so have a career talk, to learn
  what to do.
Share your interests and goals, and ask questions too,
Then listen intently for tips on 'How To'.
You'll learn quite a lot, many interesting views,
That will guide you to know which direction to choose.
To learn more or contact Joanne:

Source: Adapted from Green 360 Career Catalyst, linkedin.com/in/joanne-martens-gr8coach mailto:
joanne@theintelletoadvantage.com www.green360careercatalyst.net. © John Wiley & Sons.

Company Profile

The Intelleto Advantage

The Intelleto Advantage is a leadership and career coaching and development
organisation.

  We help our clients develop meaningful careers, transition successfully into new
roles and jobs and expand their leadership potential.

  We specialise in coaching and developing leaders and teams to achieve success and
fulfillment through focused and principled leadership, conscious action, authentic
communication and effective decision-making.

  We developed Green 360 Career Catalyst, an online career preparation and environmental sustainability resource for students to make informed career decisions and create pathways toward green careers. Green 360 enhances high school education
programmes and young adult training programs. Free access available to individuals
around the world.

Source: Adapted from Green 360 Career Catalyst, linkedin.com/in/joanne-martens-gr8coach joanne@
theintelletoadvantage.com www.green360careercatalyst.net. © John Wiley & Sons.

Personal Profile

Charlotte Lin, Founder and Career and Climate Action Coach at Green Growth Coaching, Canada

As a lively and enthusiastic climate change coach, Charlotte specialises in helping change-making leaders worldwide develop green careers, carbon-neutral businesses, organisational engagement and environmental wellness. With a holistic approach, Charlotte guides changemakers to be better for themselves, their communities and the planet.

Charlotte's love for learning has been fueling her adventures around the world since she was young. Curious and creative, she has worked as an arctic ecologist, a fashion photographer, a corporate language consultant, an academic researcher, a university lecturer and a design consultant. Assimilating her vast interests and 13 years of work experiences into the dynamic profession of coaching has made her deeply grateful and ecstatic for life.

◀ **Top Tips**

1) Continuous Learning of the World of Work and How you fit in

   New environmental careers are emerging rapidly due to the global shift to a low-carbon economy. Other sectors influence changes in environmental careers as well. New opportunities that might arise from this intersectional influence. To get the most updated picture of emerging environmental careers, check career database websites to see the latest updates on Green Economy (for example: www.onetonline.org/find/green). There are often jobs that you never even considered on the list. Never assume you already know all the possible job options that are available to you. Do the groundwork and your chance of succeeding will greatly increase.

   Other than job databases, it is also important to understand industry trends such as job availability, growth, salary, education and skills needed. The technical knowledge of the world of work will give you some initial ideas on what is possible and what you might pursue. There are also numerous career assessments that can help you decide what is a good fit for you.

   The other aspect of this tip is how you, as a whole person, fit into the world of work. Essentially, this comes down to understanding yourself, including things like your personality, your interests, your strengths, your expertise, your values, your likes and dislikes, your weaknesses and anything else that makes you the unique person that you are. All this information will inform you about what you want, but more importantly, what you don't want. Incorporating a thorough understanding of yourself into your career planning helps to ensure that you are an integral part of your career development, which has an astonishing effect on your long-term well-being and happiness in your career.

   Most climate activists and environmental professionals I have spoken to suggest that education is key to understanding what to do about climate change.

Without the necessary knowledge, it is nearly impossible to take actions that are coherent with your goals. The ability to learn autonomously and independently ensures that you will always be able to find the necessary knowledge to move forward.

Once you dive into learning all about the environment, though, it is easy to become overwhelmed with information. A great way to not be overwhelmed is to be able to effectively organise the useful information based on your goals. Identifying a big-picture goal will help you anchor both your learning and organisation.

Identifying your big-picture goal requires a sound understanding of both yourself and the world of work. Figuring out which is missing will help you move forward with ease.

2) Maintaining Well-being, Hope and Resilience with Nature

Understanding yourself ties in strongly with your career well-being, which ties into your overall well-being and happiness. Well-being, in my opinion, is more important than pretty much anything else I will talk about. This is because if you are not happy, it is very hard to implement the knowledge and execute the steps necessary to develop your career in the way that you truly want. Well-being includes aspects of our lives such as sleep, exercise, nutrition and finances. Problems in any of these areas can get in the way of your career development and take you further away from your goal, which can cause further discouragement.

This negative cycle has a significant impact on our hope for our future. Hope is 'the perceived capability to derive pathways to desired goals and motivate oneself via agency thinking to use those pathways' (Snyder 2002). Without hope, it is easy to think nothing matters and subsequently not take necessary actions towards your goals. A lack of hope might make you give up pursuing a career prematurely or settle in under-employment or a position that you're not happy with.

One way to stay hopeful when facing challenges is to develop your resilience. A resilient person exhibits the ability to quickly recover from adversity and adapt to new situations with positive actions (Block and Kremen 1996). In the case of environmental careers, hope and resiliency not only influence our thinking regarding our career goals but also our goals and dreams for the planet. With the larger goal of protecting our planet, we can't afford to lose hope in our pursuit of environmental careers. It is important to stay resilient when times get hard.

One easy way to restore our resilience is spending time in nature. Research has shown that nature has restorative effects on our attention capacity for the elderly (Ottosson and Grahn 2005), aggressive behaviours in young adults (Kuo and Sullivan 2001), and mental fatigue (Berto 2005). It also makes sense that if you want to work to protect nature, you should spend time to really immerse yourself in it. Connect yourself deeply to it. We already live in an age of separation, where humans formed such a deep cavern between our built and natural environments. It's not a narrative that is easy to unlearn and overcome. Take your time to understand all parts of nature, a massive and intricately connected system with millions of inhabitants. What is your relationship with nature? What does nature mean to you? What have you learned from nature? How is nature connected to your work and life? What is your mission?

Working on something bigger than yourself can feel daunting. A deep, personal connection to nature will inspire empowering and intrinsic motivation for you to stay resilient in the face of challenges. Gratitude and love of nature will provide genuine drive to protect it. Nature will also revive you when you feel the stress from our human-built environment. Nurture nature and it will nurture you back.

3) Thinking Outside of the Box and Prototyping your Ideas

As we shift towards the low-carbon economy, the way we work now will shift as well. This means new sectors, new methods, new leadership, new ways of collaborating and thinking are emerging every day. Creativity is the key that gives us that flexibility to adapt in an unpredictable world.

In addition, the low-carbon shift is the perfect time for us to dream up a new future that is sustainable – a time to really think about what has worked and what hasn't, and where do we collectively want to go from here. Coming up with a new story about how we operate on this planet takes creativity, imagination, and courage, but it will serve as the foundation for your new reality – so dare to dream, be bold, and claim the future you want.

Don't let the old definition of environmental careers limit you. Even now, there are many jobs that don't immediately seem environmental-related but are actually connected in unexpected ways. In fact, I would argue that all jobs can be related to the environment as long as we are willing to explore the possibilities. Start with your own unique skills, talents, strengths, points of view and how you want to contribute to the environment, you will find where you belong.

But what if your new-found possibilities become too overwhelming and paralyzing for you to make a decision? How do you know which path is the right one for you? A lot of times, a dream is very different from the reality. This is where prototyping comes in – a concept from the world of design and engineering, where a simple model of an idea is built out and tested before further development. Prototyping helps you avoid huge time, energy or monetary commitments before you decide to jump into something. Prototyping is coming up with a quick way of actually experiencing your idea so you can make a decision on how to proceed.

Decision making is not unique to environmental careers. Decision making also doesn't only happen at the beginning of your career. Choosing a path to pursue is only the first of millions of decisions you have to make throughout your career. Getting good at decision making is going to make your life and career a lot easier.

4) 'Networking' – Leadership, Collaboration and Building a Community

Many environmental careers are about making the world a better place, which goes beyond personal gains and advancement. So, what are the bigger goals that you are serving?

We so often think career development is just about the individual, but it's not. It's about what values you can bring to others and the world. What the world really needs right is good leadership and systemic change, which ultimately requires you to be good at working with others from a place of selflessness and service.

What might networking look like from a place of selflessness and service? No matter what stage of career you are in – networking and building a community is important for your career wellbeing and development. My preferred way of 'networking' is a big-picture approach where you focus on building a coherent personal brand with a solid mission.

Personal branding is not just for entrepreneurs or businesses – everyone has a brand, even if you don't think about it that way. Your 'brand' will naturally influence the way you attract, reach out to, and interact with like-minded people. Networking, in this context, is just authentic and enthusiastic interaction based on your genuine passion and interest. Figuring out your personal brand starts from understanding yourself.

Personal branding should be developed while you undertake solid actions to learn about environmental careers – things such as internships, volunteering, extracurricular activities, improving communication skills on and offline, joining local chapters or groups for environmental activism or careers, contacting university or community career centers, and more. Don't stop networking even when you are studying or at another job. Having a purpose or a mission in mind (these are part of your personal brand) will help motivate you and anchor your actions in the long run. In addition, don't just network for your resume or other personal gains – always try to offer something in return. What can you offer in return? Dig into your uniqueness and you'll see many ways you can contribute. In the spirit of learning, have a curious conversation with someone you're interested in. Ask about them and their stories. Ask about anything – things you don't know and things you do know. It's always great to have a different perspective.

Finally, and perhaps more importantly, your community will become your support system on the often difficult road of an environmental career or the activism that you may participate along the way. Burn-out and disillusionment are common in environmental careers for many reasons. The systemic changes you long to see from your hard work may take years to happen – or never at all in the timeline of your career. While there is no quick solution for this, being surrounded by the right people as you work towards your goal is absolutely crucial.

5) Working with a Career Coach

A career coach acts as a trusted partner, a listener, a cheerleader, a truthsayer and a guide to help you navigate through all the points below. Your coach can help you identify your goal and plan strategic steps towards it. As you move towards your goal, your coach can keep you motivated and resilient when times are hard, by helping you overcome any problems that might keep you stuck. Your coach is also a neutral voice that will provide you unbiased guidance that your friends and family may not be able to do. From an unbiased point of view, a coach could also effectively lead you to discover new options that you may have never thought of before by yourself.

If you're starting fresh in an environmental career, a coach can also help you rebrand yourself and make sure all your past experiences and transferable skills are presented in the best light possible. Together, you can also plan strategic and intentional actions to establish your personal brand and network as you advance in your career. Most importantly, whatever it is that you want to achieve, your coach can keep you accountable to the actions you decide to take, so that you can move forward in the right direction with strong momentum.

You can follow Charlotte's work at www.greengrowthcoaching.com

Source: Charlotte Lin, Founder & Career and Climate Action Coach at Green Growth Coaching, Canada. © John Wiley & Sons.

Personal Profile

Janavi Da Silva, MSc Pr.Sci.Nat, Programme Manager, GreenMatterZA, South Africa

I am the Director of Programmes at GreenMatter, the engine for growing the minds that shape the green economy in South Africa. I am responsible for implementation of Human Capacity Development programmes across the organisation. I have a deep curiosity for social and environmental enterprise and skills development, and I believe in working from someone's highest values and creating a unique place in the market for their special skillset and values. I coach green professionals and green entrepreneurs to make their greatest impact. I aim to help people to participate meaningfully in the Green Economy on a global scale.

I have an Environmental background and an interest in working towards social change. As a result, I studied a Master's degree in Environmental Science at the University of Johannesburg and have a Diploma in Social Entrepreneurship from GIBS Business School. I am also a facilitator and digital marketing enthusiast!

◄ Top Tips

1) A deep awareness and a clear picture of your personal mission statement and your true values will lead to your greatest legacy.
2) Volunteerism is also a great way to get noticed and be found in the green economy, even if you just volunteer on a project basis.
3) The best way to add real value in your career, towards promoting a sustainable future, is to work in a few disciplines that are opposing but complimentary. Say, a Marine Biologist who is also a circular economy expert. These two disciplines go hand in hand, because the marine ecosystems could do with more circular economy designs to prevent plastic waste from entering the ocean. These are rare pairings of skills but are in dire need to make real changes
4) I also strongly believe in developing your digital skills in a changing world and the best way to reach people now and more so into the future will be through the internet in some way or another. Figure out which digital skill you can use to really stand out from the crowd and create a lasting impact in your field.

Source: Janavi Da Silva, MSc Pr.Sci.Nat, Programme Manager, GreenMatterZA, South Africa. www.linkedin.com/in/janavi-da-silva-pr-sci-nat-a90409aa. © John Wiley & Sons.

Organisation Profile

GreenMatter

'Greening the minds that shape our Planet's future'

GreenMatterZA is the engine for growing Biodiversity skills. Developing the right people at the right time for the green economy.

GreenMatter is an initiative that drives transformation in graduate level skills for Biodiversity. Co-founding partners SANBI (the South African National Biodiversity Institute) and the Lewis Foundation led the development of a Biodiversity Human Capital Development Strategy (BHCDS) in 2009–2010. The strategy is aimed at producing specialist, research and management skills for all organisations with biodiversity mandates, objectives, interest or impact, in the context of skills shortages, the need for social transformation and opportunities for growth and employment in the greening of the South African economy.

GreenMatter activates through the involvement of a range of organisations, institutions and partners (which include NGOs, SMME's, parastatals, national and provincial government departments), Higher Education Institutes (HEIs), SETA's and Business, in delivering through a shared implementation model on the needs for quality skills and transformation.

Source: Adapted from GreenMatter, www.greenmatterza.com. © John Wiley & Sons.

Personal Profile

Megan Fraser, Certified Coach, UK

I began my career in start-up consultancy in Canada, where I became fascinated by how effective leadership impacts an organisation's success (or not). This led me to start studying coaching in 2015. I'm now a certified professional coach and facilitator based in Scotland and I love working at the intersection between personal development and wider societal change.

In my private coaching practice, I work 1–1 with professionals, leaders and entrepreneurs who want to create a life and career in line with their values. Often my clients come to me in transition, feeling stuck or lost as they try to figure out next steps. Many of my clients have a deep sense of social responsibility or an awareness of the climate change emergency – whether or not they're currently working in the green economy – and want some help to decide how to use their talents to contribute meaningfully. I'm also a member of Climate Change Coaches, where we see climate change as a human behaviour problem, and help people take action around it.

I've coached at The Journey, the world's biggest climate innovation summer school, where we help young people think about how we might tackle climate change through entrepreneurial thinking. I also coach at Good Ideas, Scotland's incubator for social innovation; and at The Lens; a social enterprise that develops leadership and intrapreneurial thinking across the Third and Public Sectors in the UK and beyond.

◄ **Top Tips**

Green Job Searching Tips:

1) Create your own structure and goals for active job searching. Without this, it can feel overwhelming and very difficult to quantify your success. It's helpful to adopt a simple mantra of 'one seed a day'. Importantly, this 'seed' needs to be an action in the *outside* world: time spent polishing your resume, updating your LinkedIn profile, or researching jobs don't count as seeds (though all those things are still part of your search).

   A seed might be an informational interview, a job application, an interview, an email to someone requesting a phone call, a message to your network asking for introductions to particular types of professionals. Once your seed has been planted, you've done all you can do and can move onto the next seed. Of course, you can plant more than one a day, but one is your minimum.

2) Don't mistake quantity for quality. I've met many people who tell me they've applied for 'hundreds' of jobs online and haven't heard back from any of them. The reality, however, is that sites like Indeed and LinkedIn – while allowing us to technically 'apply' via one click – can be black holes for resumes. And if your resume isn't optimised for their algorithm, it might never even be seen by a human. Instead, focus on good quality networking (below). One solid informational interview is worth innumerable one-click 'applications'. Don't chase the numbers.

3) Be proactive in 1–1 networking. Surveys indicate that between 60 and 85% of jobs are found through relationships, not job boards. Research organisations you're interested in working with and follow them online so that you're up to date with their news. Read environmental blogs and magazines to learn who the main players are in fields that interest you. Actively start finding out who you already know in those organisations, or who might be contacts of people you know already.

   Start inviting these people for 30-minute informational interviews where you can learn about their jobs and whether you can see yourself growing into them. Always ask them for a referral to someone else, too! Often, they'll introduce you to a couple of people, each of whom you can then interview too. Your network of quality contacts will steadily grow. Stay in touch with everyone you meet – but don't be pushy or try to sell yourself. Aim to help *them*. Be friendly. This will help you stay on their radar.

4) Always over-prepare for interviews. Never make the mistake of 'winging it'. An interview is the fruit of many hours of research, networking and hard work. This is not the time to finally relax. This is the time to bring your best game!

   Find good 25–30 sample interview questions online and take the time to actually write out your responses to these. You don't have to remember your answers word-for-word but writing them will help you know exactly what you want to say. Then rehearse them repeatedly in front of a mirror and with a friend. Get feedback about your responses (and whether you're waffling) and tweak your responses until you sound fluid and are answering exactly what the questioner is asking.

5) Lastly: if you're drawn towards part- or full-time entrepreneurship, remember that even purpose-driven organisations need to be business literate. It's not enough to be passionate about solving a problem. Often social innovators are so driven to fix something, they neglect the need to plan for revenue streams. And at times, the problem they're trying to solve (like climate change) is so huge that they don't think strategically enough about how to tackle just one particular part of it.

In your planning, always put the customer (or user) first. If they don't see value in what you provide, there's no business. The Strategyzer tools, available online for free, help you map out your value proposition and potential business models and will give you lots of additional tools to grow your ideas.

Source: Megan Fraser, Certified Coach, UK. www.meganfraser.org www.linkedin.com/in/meganclarefraser. © John Wiley & Sons.

Top Tips

Zoe Greenwood, Oxford HR Consultant and Climate Change Coach, UK

1) If you are passionate about the environment but you're not a scientist, engineer or something more conventionally 'suitable' for the sector, then remember that you can use your skills in service of the environment whatever they are. Many of the organisations listed in this book need fundraisers, HR, finance, communications and more. Journalists, lawyers, educators and coaches can all specialise in the environment. In summary, to meet the UN Sustainable Development Goals, we need people with a wide cross section of skills and competencies. Whatever your professional interest is, there will be a space for it.

2) This is a sector that is changing by the day and with any luck fast becoming business as usual. Climate change has gone mainstream, businesses no longer want to be compliant but sustainable with transparent supply chains, coastal cities the world over will need to adapt to rising sea levels and growing our own food is trendy. If you expect to be an expert, you will be waiting a long time – instead, get stuck in and remember that what is considered best practice today may not be tomorrow. Be prepared to keep learning and asking questions.

3) Think about becoming a trustee of a not-for-profit organisation connected to your area of interest. Increasingly NGOs are looking for young people to join their Boards and seeking trustees from different industry backgrounds. This is a fantastic way to learn, contribute, get noticed and stay motivated.

4) Make the most of our interconnected and virtual world to find people who can share tips, ideas and experiences. Most people who are further on in their careers are happy to share their stories with people starting out or looking for a career transition. Networking can open doors and no longer needs to be face-to-face.

5) A career in the environmental sector can be tough at times; after all, despite great wins we are still losing the battle. It is okay to feel frustrated and sad. Make a plan to support your own resilience – being outside in nature is a great reminder of why we do what we do (and not all environmental jobs are outside, in fact many of them are desk based), find yourself a support network or an 'environmentally friendly' coach to keep buoyant. Follow your passions and never doubt that what you're doing is important.

Personal Profile

Zoe Greenwood co-leads the Environment, Climate and Conservation practice area for Oxford HR, an executive search firm who find and support world changing leaders. She is also a founding member of the Climate Change Coaches. Before this, Zoe spent 15 years working for NGOs.

After completing her degree (not related at all to the environment!), she travelled to Hong Kong and Indonesia and supported the delivery of environmental education programmes in schools. She went on to run the Press Office for Earthwatch before retraining as a coach and joining the Learning and Development team, working with multinational companies to deliver sustainability leadership journeys. These roles took her all over the world – witnessing the natural wonders of our planet and the scale of degradation.

Source: Zoe Greenwood, Oxford HR Consultant and Climate Change Coach, UK. www.linkedin.com/in/zoegreenwood. © John Wiley & Sons.

Personal Profile

Shannon Houde, Career, Executive, and EI Coach, Walk of Life Coaching, UK/USA

I've been an eco-entrepreneur, a management consultant, an MBA student, and a multilingual traveller. My 20-year career story has been one of constant reinventions, so I know first-hand how to help others make their career-change dreams a reality. I chose to create my dream job in career coaching to combine my diverse experience as a hiring manager, business coach, Corporate Responsibility management consultant, and corporate recruiter.

I coach leaders to shift their mindsets about what is possible across three stages of their careers: finding purpose, growing influence and delivering impact. I provide emotional intelligence coaching and design team coaching courses around diversity, equity and inclusion. I have coached more than 1000 clients in more than 40 countries and I love helping people and organisations achieve their potential for impact by connecting to their purpose.

It wasn't until I founded Walk of Life Coaching that I realised I had discovered my true calling. At the intersection of recruitment, matchmaking, personal branding, thought leadership and responsible business, I found my sweet spot where I get the right people into the right roles to create a more sustainable world.

Top Tips
- *Get the lay of the land.* The purpose-driven economy is forever evolving and the jobs market right along with it. Make it your business to read and research as much as you can about the work you hope to do and the players in that space. Cultivate a deep understanding of the issues at play, because environmental work is increasingly nuanced and sophisticated.
- *Identify your values and traits.* Zero in on what guides how you behave and what has shaped your life. This will help inform the language you use in composing your personal brand and how you develop a compelling narrative about who you are and why you are valuable to potential employers. It will also help you to align your

core values to the types of organisations you pursue for employment and to determine what attributes of an employer and position are non-negotiable for you.

- *Identify your superpower.* There are few direct paths to green careers. Many leaders in this space started out in roles that had nothing to do with it – such as in engineering, marketing, or operations, among others. Before you determine which route to take, you should identify your 'superpower' skills, which can help set you apart. This can be anything from strategy to communication to project management to programming.
- *Talk about your accomplishments, not your past positions.* Providing a laundry list of your previous roles and employers may seem like the default approach in job applications, but it's not especially useful to hiring managers. Instead of framing things in terms of what you were tasked with doing, present your background in terms of what you did do and what impact it had. Talk about the results you have delivered in concrete terms. Did you help remediate former brownfields? Secure funding for a habitat protection project? Develop a software that monitors the migration of ocean debris? These are the types of details that really communicate what you have to offer.
- *Cultivate your emotional intelligence.* Effective sustainability leaders align their intellectual intelligence with their emotional intelligence, and organisations are waking up to that fact. Just look at all the mindfulness courses that major companies like Google are offering their employees. They understand the importance of being able to keep your ego in check and empathise with other people, especially when things are fraught, and solutions seem hard to find. Emotional intelligence is also connected to the ability to coordinate with others and manage teams of people, which are crucial skills when working towards an environmental mission.

Shannon Houde is an ICF-certified career and executive coach who founded Walk of Life Coaching, the first international career coaching and professional development advisory business focused solely on the social impact, environmental, CSR, and sustainable business fields. Get in touch to join our wider network or to book a coaching session.

Source: Shannon Houde, Career, Executive, and EI Coach, Walk of Life Coaching, UK/USA. www.walkoflifecoaching.com. © John Wiley & Sons.

Personal Profile

Catriona Horey, Leadership & Life Coach and Nature & Climate Change Coach, UK

I grew up gently immersed in the environmental world. My father, Geoffrey Matthews, was one of the founding fathers of the Ramsar Convention on Wetlands and spent more than 30 years as Director of Research and Conservation at WWT (Wildfowl & Wetlands Trust), headquartered in Slimbridge, where he worked closely alongside Sir Peter Scott. My mother, Mary Matthews (née Evans), also worked at WWT, conducting research into swan behaviour over a decade.

I grew up in the Gloucestershire countryside and, as a child and teenager, would write letters to MPs about environmental issues, was a member of green charities and raised money for environmental initiatives.

My interest in the environment continued into my late teens and early 20s. I collected data for a GTZ project on fire management and prevention in the grasslands and forests of Mongolia, as part of a Raleigh International expedition. My degree in Human Sciences at the University of Oxford had biological and ecological elements, and I helped at a sustainable watershed development organisation in China.

Then, I somewhat changed direction: I spent more than a decade focusing on humans! After my Master's in Humanitarian Assistance in Bochum, Germany, I worked to help transform the performance of non-profits and social enterprises and leaders in social innovation, both in London and New York. I then retrained as a certified leadership and life coach, working with leaders in social innovation and women looking to redefine their professional and personal priorities.

For the last few years, perhaps in part having become a mother and with climate change higher on the world's agenda, it feels like I'm starting to 'come home' or 'full circle' back to green issues. I now offer coaching outdoors in Battersea Park, London, as part of my packages and have co-led nature-based workshops there too. I participated in a fabulous coaching circle with Climate Change Coaches and now offer coaching to people who want to explore their relationship with climate change so they can move from scarcity, overwhelm, anger and/or grief to a place of resonant action. I have also co-delivered climate change mindset workshops. And I co-host a podcast that explores the power of the natural world to inform and inspire with Elizabeth Wainwright, a writer, a coach and consultant and an elected Green District Councillor. I'm excited to train further and dive more deeply into nature-based and nature-inspired work alongside my leadership and life coaching.

◄ Top Tips

1) It's never too early or too late to get involved in environmental or green issues.

2) Reconnect with times in your life where you felt really on purpose, empowered and in action and, from that place of resonance, consider what you want your legacy on this earth to be. Follow your intuition – even if your path might not be straight.

3) Be creative with the skills, strengths and passions you already possess and think about how you might apply them with a focus on the environment. Remember the green agenda is increasingly being woven into different industries and roles so there are many new and diverse ways to get involved.

4) Search out training, mentoring and/or coaching opportunities to help develop yourself, whatever stage in your career you are. Be curious, explore what could work for you and, for more expensive offerings, don't be afraid to ask if there are scholarships available.

5) Be bold and creative with your connections and network. My career has been built on meeting people through 'informational interviews' and referrals. Explore LinkedIn and your wider network and invite people you admire to in-person or virtual coffees to learn more about their work and find out which organisations

> they might recommend to you and people they might suggest you speak to – indeed they may have an opportunity available themselves or offer to refer you! And then pay-it-back by sharing your wisdom and connections with those coming after you. And enjoy your explorations!
>
> *Source:* Catriona Horey, Leadership & Life Coach and Nature & Climate Change Coach, UK. www. withcatriona.com. © John Wiley & Sons.

1.11   Reality Check

It is worth having a reality check in terms of working in the environment sector. There are many ways that you can benefit the environment, even if you don't work in the sector. Many people volunteer outside work for NGOs or make donations to contribute to green efforts.

The excellent 2014 blog article '5 ways to make a big difference in any career' by Benjamin Todd (2014) on the 80,000 Hours website notes:

> At 80,000 Hours, we're focused on finding the very best opportunities for you to do good with your career. We're worried that sometimes this continuous focus can be demoralising. After all, it's hard to find the best opportunities. Moreover, we're worried that sometimes our members lose sight of the fact that *you can make a big difference in any career.*

A 2015 BBC article bluntly notes 'it could well be effectively better to become a very rich banker and give away your money than go and work for an NGO in Kenya.'

A common route taken to a green career is being an environmental consultant, but as EarthHow' (2021) highlighted in the article 'The Good, the Bad and the Ugly for Environmental Consulting Careers':

> After graduation, you'll probably want to land a job where you can make a positive difference to the environment. The startling truth is that 'making a difference' is not the job description of an environment consultant. Your very existence impacts the environment by documenting local conditions prior to human impacts, evaluating human impacts, or cleaning up human impacts. You are a scientist or engineer collecting data, developing a solution and writing reports. You enforce environmental regulations. There are rare opportunities for you to change the world.

Within the environment sector, pay is generally not high, especially within NGOs. However, competition for roles is very intense for most roles.

The importance of certain personal qualities has already been mentioned in the book introduction. There are some personal qualities that are essential for someone seeking to develop an environmental career. As for many roles, there are skills that you should be able to demonstrate – teamwork, negotiation, languages, and good time keeping. One word commonly mentioned is 'stickability'. You need perseverance to do well in a sector which is very popular and where roles can be poorly paid.

There is a major stumbling block for those wanting to begin a career in the sector – the vicious circle of 'no experience no job, no job no work experience'. There are a number of sectors where you now have to pay for work experience to enable you to gain professional development and transferable skills. This vicious circle can be a thorny problem unless you have the resources to afford 'paid for work experience'.

There has been a backlash against this practice and against the offering of unpaid work experiences which are in practice using free workers to undertake tasks that would normally be paid for. There is no easy answer to this challenge. I carried out a lot of volunteer work whilst at school and I was fortunate enough to have a paid work experience at university. The earlier book section on Volunteering noted the pitfalls of volunteering and internships.

In terms of training, there are many options for increasing your qualifications and skill-sets. Some training providers are costly (and some of these courses are worth the cost), but there are many courses delivered by local, national, and online providers, which have a low cost or are free.

Mental Health Issues

There are significant issues within the green sector in terms of mental health. The transition to a greener economy inevitably involves change and this in return can increase the prevalence of mental health issues within workers in the economy as well as the general public.

The European Agency for Safety and Health at Work, in its (2013) report 'Green jobs and occupational safety and health: Foresight on new and emerging risks associated with new technologies by 2020' highlighted some key issues developing from increasing uncertainty in a changing economy:

- Economic growth concerns around the availability of funding for green initiatives
- Green values and the willingness (or not) of the public to change
- Challenges with innovation and new technologies

The report summarised:

> During this work, it became apparent that many of the current assumptions about future green jobs are based on an optimistic outcome, a Win-Win scenario. But the possibility that these targets are not met should be taken into account.
>
> As diverse as green jobs may be, a number of common challenges were highlighted:
>
> - Decentralised work processes: as workplaces are getting more dispersed and more difficult to reach, monitoring and enforcement of good OSH [occupational safety and health] conditions and safe working practices is likely to become more challenging
> - A growing use of subcontracted work, as well as an increase in self-employment, micro and small enterprises: such structures may have less awareness of OSH and a less-developed culture of OSH, as well as fewer resources available for OSH and less access to OSH services

- New skills and the need for adequate worker training: there are many new green technologies and working processes where specific knowledge is needed but has not yet been fully developed; there are also (new combinations of) 'old' risks but found in new situations equally requiring new (combinations of) specific skills; the job opportunities in green jobs may attract new entrants extending beyond their original skills areas and unaware of these new challenges
- Skill shortages and polarisation of the workforce, with low-skilled workers pushed to accept poorer working conditions and more difficult jobs
- Increased automation, which may improve OSH but also bring human-machine interface issues as well as issues of over-reliance on the technology
- Conflicts between green objectives and OSH, with the risk of OSH being overlooked
- Novel, difficult-to-characterise and potentially hazardous materials that will need to be closely monitored over their entire life cycle for potential (unknown, long-latency) health hazards: this will be increasingly challenging as no one stays in the same job for life, making it difficult to link health effects to occupational exposure.

Some areas of work within the green sector have high levels of stress and confrontation. These including International Development work and Political Lobbying. In addition, there is considerable pressure in committing yourself to a career that aims to make a difference in the world and the potential stress when you review whether you have achieved as much as you wanted.

Conclusion

A green career can be a life-defining experience, and working in an environmental job can be hugely rewarding – often you can be directly improving your local environment to the benefit of current and future generations.

However, you need to have a reality check before deciding that a green career is your aim. Issues of intense competition for work experience, let alone for low paid jobs, and the challenge that in some roles you may feel that you are not making 'enough' change to the world may weigh on you.

1.12 Social Media and Online – Popular Resources

There are many 'catch all' and more focused sites and resources that cover more than one sector. In recent years, the growth of online content websites like YouTube and social media platforms including Facebook have enriched the information available to job seekers.

Developing websites is not cheap and content is relatively inflexible, whereas on social media platforms you can form groups and add content easily. Consequently, many areas that are under-represented on websites are well covered on other social media such as Facebook, LinkedIn and YouTube – issues such as local green work, events, global case studies, small projects and very focused topics. Another highlight is that you can more easily engage with other green professionals using social media.

There is also a significant growth of events, mentoring, interviewing, virtual networking and interaction using low-cost videoconferencing platforms and social media. This helps to make communications more efficient and global.

There is a growing number of jobs websites and specialised recruitment agencies that focus on green careers.

Some of these resources are listed below, but please note that this is a very small selection of the content which may be helpful in your career development.

Box 1.4   Multi-Sector Resources and Specialised Jobs Sites

Facebook Groups and Pages

All International Environmental Sciences Jobs and News
www.facebook.com/groups/1513496342234562

Environmental Career Experts
www.facebook.com/EnvironmentalCareerExperts

Environmental Career Opportunities
www.facebook.com/ecojobsource

Environmental Education Career Network
www.facebook.com/groups/1625744117557119

Environmentalists in Pakistan!
www.facebook.com/groups/environmentalists.pak

Environmental Job Board & Forum
www.facebook.com/groups/envjob

Environmental Job Opportunites SEA: Internships/Volunteers/Learning
www.facebook.com/groups/130704717461155

Environmental Jobs
www.facebook.com/EnvironmentaUobsUK

EnvironmentaUobs.com
www.facebook.com/EnvironmentaUobs

Environmental Jobs Network
www.facebook.com/EnvironmentaUN

Environmental Jobs and Free Career Resources
www.facebook.com/groups/1297214433774855

Environmental Management university group
www.facebook.com/groups/201806079916217

Environmental Professionals
www.facebook.com/groups/412035932889946

Environment Jobs – Bulletin Board
www.facebook.com/groups/envjobs/

Green Careers DC
www.facebook.com/groups/223689644750214

Green Careers in India
www.facebook.com/groups/419579064751703

Green Jobs (UK)
www.facebook.com/GreenJobsUK

Greenjobs (USA)
www.facebook.com/greenjobscom

Greenjobs (Netherlands)
www.facebook.com/greenjobs.international

Green Jobs Network
www.facebook.com/groups/greencareers

Open – Wildlife and Fisheries Job Board
www.facebook.com/groups/791215834226997

Sustainable Agriculture Jobs, Internships and Apprenticeships
www.facebook.com/groups/SustainableAgJobs

Wildlife Science Career Network
www.facebook.com/groups/457674185029274

Wildlife Workers' Network
www.facebook.com/groups/wildwork

LinkedIn Groups
Australia Green Jobs & Career Network
Conservation Jobs – Careers & Talent Network
Environmental Careers Network
Environmental jobs, conservation jobs, green jobs, energy jobs, sustainable development, GoodWork.ca
Green Jobs & Career Network
Jobs in Nonprofits
Sustainability Career Group
Green Career Path

YouTube content
Countdown – a global initiative to accelerate solutions to the climate crisis
www.ted.com/series/countdown

14 Exciting Environmental Careers that Make a Difference
www.youtube.com/watch?v=QmAAEoc6_sE

10 Environmental science careers you should know about (& salaries!)
www.youtube.com/watch?v=iFp7QujrwS0

10+ Wildlife biology careers you should know about (& salaries)
www.youtube.com/watch?v=AK2Onqx-ctM
   (Kristina Lynn www.youtube.com/c/VeganBelowZero/featured)

Where to Find Wildlife Work
www.youtube.com/watch?v=suGoLESaE60

Stephanie Martin

Multi Sectoral Websites
O*Net Online – career outlines across sectors
www.onetonline.org

80,000 Hours
www.80000hrs.org

Green Drinks – Every month, people who work in the environmental field meet up at
   informal sessions known as Green Drinks. Now in more than 300 cities worldwide.
www.greendrinks.org

Goodwork
www.goodwork.ca

Canadian green jobs site
EnvironmentalCareer.com - lists environmental job openings in private and public interest
   settings. Search by location, field, area of substantive expertise and other criteria. You
   can also post your resume for employers to view online. Career coaching is available.
www.environmentalcareer.com

Partners Achieving Sustainability Excellence (PASE) Corps – a non-profit organisation
   for research, training and on-the-job experience in environmental conservation and
   sustainability initiatives.
www.pasecorps.org

Cyber Sierra's Natural Resources Job Search lists jobs in conservation fields and provides pages with links to other environmental job search websites.
www.cyber-sierra.com/nrjobs/index.html

National Association of Environmental Professionals
NAEP is an interdisciplinary organization dedicated to developing the highest standards of ethics and proficiency in the environmental professions.
www.naep.org

Environmental Careers Organisation (ECO) Canada
ECO Canada is a steward for the Canadian environmental workforce across all industries.
www.eco.ca/about-us
www.eco.ca/training

APS jobs – Australian Public Service and government jobs
Central government jobs portal
www.apsjobs.gov.au

Centre for Alternative Technology
www.cat.org.uk/courses-and-training

Incubators
These incubators – or labs – support young people to realise their goals. They provide fellowships, grants, mentoring support, resources, networking and physical space to students who are driven to realise social innovation.

The School of Visual Arts Design for Social Innovation
www.dsi.sva.edu/apply
www.dsi.sva.edu/partners

Net Impact
www.netimpact.org

The Asia-Pacific Youth Network, Green Jobs Network
www.apgreenjobs.ilo.org

YouthBuild
www.youthbuild.org

FOR UPDATED AND ADDITIONAL RESOURCES, INCLUDING EXTRA CHAPTERS, GO TO WWW.ENV.CAREERS

References

BBC News (2015) What's the best way for Zuckerberg to give away his billions? www.bbc.co.uk/news/business-34982233 (accessed 22 September 2020).

Bell, S. (2015). Volunteer tourism and conservation: effects and consequences of motivations (accessed 2 April 2021). www.researchgate.net/publication/327727812_Volunteer_Tourism_and_Conservation_Effects_and_Consequences_of_Motivations

Block, J. and Kremen, A.M. (1996). IQ and ego-resiliency: conceptual and empirical connections and separateness. *Journal of Personality and Social Psychology* 70: 349–361.

Rita Berto, Exposure to restorative environments helps restore attentional capacity, *Journal of Environmental Psychology*, Volume 25, Issue 3, 2005, Pages 249–259, ISSN 0272-4944, https://doi.org/10.1016/j.jenvp.2005.07.001.

Brookings Institution (2011). Sizing the clean economy: a national and regional green jobs assessment. www.brookings.edu/research/sizing-the-clean-economy-a-national-and-regional-green-jobs-assessment (accessed 2 April 2021)

Cambridge Dictionary Website (2021). Cambridge University Press. www.dictionary. cambridge.org/dictionary/english/mentor (accessed 2 April 2021).

Career Action Center (1997). Career self reliance, a process for lifelong learning and development.

EarthHow (2021). 'The Good, the Bad and the Ugly for Environmental Consulting Careers. https://earthhow.com/environmental-consulting-career/

Environmental Careers Organization (1999). The Complete Guide to Environmental Careers in the 21st Century. Island Press.

European Agency for Safety and Health at Work (2013). Green jobs and occupational safety and health: foresight on new and emerging risks associated with new technologies by 2020'

Forbes (2013). 10 things you should know about career coaching. www.forbes.com/sites/ learnvest/2013/07/09/10-things-you-should-know-about-career-coaching (accessed 28 August 2020).

Hance, J., The Guardian (2017). All work, no pay: the plight of young conservationists. www.theguardian.com/environment/2017/aug/17/all-work-no-pay-the-plight-of-young-conservationists (accessed 2 April 2021).

Handwerk, B., National Geographic Blog (2012). Just what is a green job anyway? www.blog. nationalgeographic.org/2012/06/07/just-what-is-a-green-job-anyway (accessed 2 April 2021).

International Labour Organisation (ILO) (2016). What is a green job? www.ilo.org/global/ topics/green-jobs/news/WCMS_220248/lang--en/index.htm (accessed 2 April 2021).

International Labour Organisation (ILO) (2018). Greening with Jobs – World Employment and Social Outlook 2018. www.ilo.org/weso-greening (accessed 2 April 2021).

Kreutzer, D., The Heritage Foundation (2012). BLS green jobs report: less than meets the eye. www.heritage.org/jobs-and-labor/report/bls-green-jobs-report-less-meets-the-eye (accessed 2 April 2021).

Kuo, F.E. and Sullivan, W.C. (2001). Environment and crime in the inner city: does vegetation reduce crime? *Environment and Behavior* 33: 343–367.

Martens, J. (2020). The Green World of Work personal correspondence. Figure 1.1 copyright Joanne Martens, 2020.

Muro, M. (2012). Green jobs: yes we can count them!. www.brookings.edu/blog/the-avenue/2012/03/26/green-jobs-yes-we-can-count-them (accessed 2 April 2021).

Natural England (2019). Monitor of engagement with the natural environment – the national survey on people and the natural environment: Headline report 2019. https://www.gov.uk/ government/collections/monitor-of-engagement-with-the-natural-environment-survey-purpose-and-results (accessed 2 April 2021).

Organisation for Economic Co-operation and Development (OECD)/Martinez-Fernandez, C., Hinojosa, C., Miranda, G. (2010). Green jobs and skills: the local labour market implications of addressing climate change. February 2010, working document. www.oecd.org/ employment/leed/44683169.pdf (accessed 2 April 2021).

Ottosson, J. and Grahn, P. (2005). A comparison of leisure time spent in a garden with leisure time spent indoors: on measures of restoration in residents in geriatric care. *Landscape Research* 30: 23–55.

Policy Exchange (2017). The two sides of diversity. www.policyexchange.org.uk/publication/ the-two-sides-of-diversity (accessed 2 April 2021).

Santander Press Release (n.d.). The future of employment is green. www.santander.com/en/ press-room/dp/the-future-of-employment-is-green (accessed 2 April 2021).

Snyder, C.R. (2002). Hope theory: rainbows in the mind. *Psychological Inquiry* 13 (4): 249–275.

South China Morning Post (2018). China has new three-year plan to clean up environment, minister says: 18 March 2018. www.scmp.com/news/china/policies-politics/article/2137666/china-has-new-three-year-plan-clean-environment (accessed 2 April 2021).

Taylor, D.E. (2015). Gender and racial diversity in environmental organizations: uneven accomplishments and cause for concern. *Environmental Justice* 8: 165–180. https://doi.org/10.1089/env.2015.0018 Published in Volume: 8 Issue 5: October 22, 2015.

Todd, B. (2014). 5 ways to make a big difference in any career. www.80000hours.org/2014/02/5-ways-to-make-a-big-difference-in-any-career/ (accessed 2 April 2021).

United Nations Environment Programme (UNEP) Report (2008). Green jobs: towards decent work in a sustainable, low-carbon world. www.ilo.org/global/topics/green-jobs/publications/WCMS_158727/lang--en/index.htm (accessed 2 April 2021).

U.S. Department of Labor Bureau of Labor Statistics (BLS) (2013). Green Jobs. www.bls.gov/green/overview.htm (accessed 2 April 2021).

2

Environmental Consultancy

2.1   Sector Outline

The website www.allaboutcareers.com (2004–2020) notes that 'Environmental consultants provide expert assessment and advisory services for their clients on matters pertaining to the management of environmental issues'. The environmental consultancy sector has demonstrated rapid development over recent decades in terms of services offered, company growth and market maturation.

The environmental consultancy sector is a significant and growing sector where some global consultancies have adapted into 'catch all' organizations with global offices, staff and contracts. There is also considerable growth in smaller, often specialised, consultancies, which can focus on single issues such as policy development, fisheries management, ecology, planning and development. In addition, there are many individual consultants who have considerable experience and high profiles, and they work with clients in specialised sectors.

The global market for environmental consulting (EC) services reached US$29.4 billion (£21.7 billion) in 2016, according to a research published by Environment Analyst (2018b). Their analysis was based on an in-depth analysis of leading international operators (loosely called the 'Global 23'), with the latest findings suggesting a much more even playing field between smaller environmental/sustainability specialists and those operating as part of larger, multidisciplinary technical and professional service providers.

There are many sectors and specialisms covered by consultancies – 'environmental consultancies' can cover not just environmental issues, but some considered less 'deep green' such as construction and power plants. Many of the biggest global companies could be seen as more hard engineering biased. However, this is a key sector to consider if you want to work in the environment sector globally. You can work for a company, which is less green, and still have a personal impact on making the company 'greener' or diversify into or focus on projects that are of more relevance. Working for a large consulting firm can also be a great stepping stone to work for companies that are more focused on green projects. See Box 2.1 for key areas of environmental consultancy work.

Global Environmental Careers: The Worldwide Green Jobs Resource, First Edition. Justin Taberham.
© 2022 John Wiley & Sons Ltd. Published 2022 by John Wiley & Sons Ltd.

Box 2.1   Some Key Areas Covered by Environmental Consultancies
Water pollution and management
Air and land contamination
Environmental impact assessment
Environmental audit, standards and compliance
Waste management
Natural resources management
Noise and vibration measurement
Project management and planning
Infrastructure development and management
International development
GIS and mapping
Transport development and management
Energy
Policy and strategy development
Mining, extractive and process industries

Different work areas ebb and flow over time in areas such as infrastructure, disaster recovery and large contracts, and market cyclicality is a key concern for multinational and multi-sector consultancies who are always seeking to plan for change and to respond swiftly. Often, larger consultancies will bring in smaller specialised consultancies to increase their capacity in focused areas.

The global consultancy Stantec notes they are 'designers, engineers, scientists and project managers', so it is a fallacy that consultancies are just global engineering consultants and contractors – the sectors consultancies work in are very varied and global in reach.

This global diversification enables consultancies to exploit new markets and develop their 'offering' in established markets yet ebb and flow into and out of markets, which vary in their income and profile. Some consultancies are known for specialisms, but this is becoming less defined as consultancies evolve and merge.

Consultancies tend to organise themselves into specific market sectors such as SNC-Lavalin who divide into infrastructure, mining and metallurgy, oil and gas and power; global consultancies are able to exploit business opportunities in many areas and across sector divides. They gather multifunctional teams for larger projects such as infrastructure development.

2.2   Issues and Trends

The key issue of changing trends and work areas within consultancies is very important for career development in the sector. There is always the risk of moving into a 'career cul-de-sac' if your sole focus is an area where companies are reducing their involvement. Again, the issue of 'generalist versus specialist' rears its ugly head. Knowing what the developing trends are within the sector is a key part of personal development.

There is a marked switch of the largest consultancies to 'mega projects' that can have a major impact on company structures and ways of working. 'Mega projects' were noted by Bent Flyvbjerg, Professor of Major Programme Management at Said Business School, Oxford University (2014) as:

> large-scale, complex ventures that typically cost a billion dollars or more, take many years to develop and build, involve multiple public and private stakeholders, are transformational, and impact millions of people.

There are many kinds of mega project, but the majority are infrastructure projects. Business Insider noted in 2017 that the world's largest mega projects 'range from giant $64 billion theme parks (Dubailand) to massive canals that will take 48 years to build (South-North Water Transfer Project in China)'.

McKinsey & Company in 2015 noted in 'Megaprojects: The good, the bad, and the better' that mega projects do have challenges in areas including over-optimism and over-complexity, poor execution, overspend and weakness in organizational design and capabilities.

Often, mega projects require consultancies to form joint venture companies (JV) to gain involvement, and this project focus presents career opportunities for those with the right skills and experience. In a special report by Pinsent Mason (n.d.), 'Joint Ventures – Delivering Global Mega Infrastructure Projects', they note:

> The successful delivery of global mega infrastructure projects will increasingly depend on bringing businesses with different skills and capabilities together through collaborative joint ventures.

Ernst & Young (2014) note in their report 'Joint Ventures for Oil and Gas Megaprojects':

> joint ventures (JVs) are now commonly used in almost all major industries. They are a key component of most major company portfolios and are seen as the solution to a number of corporate development challenges. The participants in these relationships (as in other industries) contribute assets, capital, unique expertise or labor to access diverse advantages such as scale, risk sharing, market entry, optionality, tax benefits and access to others' unique capabilities. However, while the potential advantages of cooperation are clear and well understood, ensuring these benefits are realized is not easy. . .the average JV takes 18 months to establish, yet the vast majority survives less than five years, with some research papers suggesting the failure rate for these relationships is as high as 70%.

The keynote speaker at the Environment Analyst Business Summit 2019, John Chubb (CEO, Consulting UK & Ireland) of consultant RPS, noted some of the industry 'megatrends', changes and challenges facing firms like RPS operating in the environmental consulting space:

> The exponential increase in service demand is probably the biggest change we've seen. Another significant shift is how sustainability, carbon and waste has shot up the agenda, and that has led to increasing demand from clients and regulators that we've had to respond to.

The profile of the projects that are out there has also changed – over the last 10 years we have seen many more 'supersized' major infrastructure projects like HS2 and Hinkley Point C and these projects have changed the dynamic of the sector. If you are involved in one of these supersize projects, the commitment is huge in terms of internal resource. But also, for those not involved with the big projects, the competitor landscape is very different. When firms are involved in the super projects, they often don't have the resource to go for other work. As a result, the line-up for bids can be unexpected.

There is also a focus on digital transformation within the sector. This was also noted by John Chubb:

> We recognise that technology is changing the way we work and that our people and clients are having to adapt to these changes. We know our teams want to use the newest digital age technologies to improve their ability to deliver projects to clients and make day to day working life easier.
>
> Our digital transformation strategy has been designed to maximise the combined value of our deep expertise, our data and the way we use immersive technologies to deliver unique client experiences. This includes automating away process-heavy activities and creating globally connected knowledge communities for our geospatial, BIM and VR talent, and investing in a data strategy so that in time we can generate insights about our people and our clients and be more responsive to their feedback and needs. We also know we need to invest in future skill sets, like data analysis, so our digital transformation strategy has dovetailed with our technology and people objectives to make sure we proactively recruit for data and digital acumen while also upskilling.

There is also a major trend in consultancies merging with and acquiring other companies. Examples are the 2017 merger/partnership between SNC-Lavalin and WS Atkins and Stantec acquiring MWH Global in 2016.

Many consultancies are seeing benefits in economies of scale and the possibility that a merger offers in terms of having a more comprehensive 'offering' of services and skills to clients globally. It was noted of Stantec in acquiring MWH in their press release 'Stantec to Acquire MWH, a Global Professional Services Firm with Leading Expertise in Water Resources Infrastructure':

> With the acquisition of MWH and its 6,800 worldwide employees, Stantec will gain a position as a global leader in water resources infrastructure while earning greater presence in key targeted geographies, including the United Kingdom, Australia, New Zealand, South and Central America, Europe and the Middle East.

SNC-Lavalin made a similar statement in its 2017 press release 'SNC-Lavalin completes transformative acquisition of WS Atkins':

> The acquisition of Atkins creates a global fully integrated professional services and project management company – including capital investment, consulting, design, engineering, construction, sustaining capital and operations and maintenance.

Together, we will have over 50,000 employees and annual revenues of approximately C$12 billion. This acquisition increases our customer base, geographic reach and scale, making us a true global player with more balanced revenue coverage worldwide, while strengthening our position globally to develop and capitalize on the infrastructure, rail & transit, nuclear and renewables markets.

Wood Group completed its acquisition of Amec Foster Wheeler in October 2017. Their press release 'Wood Group completes acquisition of Amec Foster Wheeler' noted:

> This transformational acquisition creates a global leader in the delivery of project, engineering and technical services to energy and industrial markets. We become a business of significant scale and enhanced capability delivering services across a broader range of geographies and sectors, differentiated by the quality of our people, enabling technology and know-how. Wood is better placed to serve customers than ever before, with a more comprehensive range of capabilities and the potential to deliver efficient integrated solutions with fewer customer interfaces.
>
> We expect to deliver significant cost synergies and incremental revenue synergies in a less cyclical business which retains a predominantly reimbursable, asset light model with a balanced risk appetite.

This process of mergers and acquisitions will clearly continue as it offers many benefits to developing consultancies. It also allows the larger consultancies to leapfrog others in the 'top 10' listings in the sector. However, there are a growing number of smaller specialised consultancies who operate independently but at times work for the larger consultancies. There is market space for specialism, as well as diversification and larger conglomerates. In terms of a long-term career in environmental consultancy, there is the likelihood that at some stage your employer name will change, or you may work as a specialist with many other consultancies as clients.

Globally, there is also a diversifying of consultancies into delivering government services and some sensitive areas such as policy and strategy development. This is slightly more complex than a privatisation of government services – governments are often utilising experts where they may not have the staff resource or expertise in-house. They may also ask consultancies to develop cross-sectoral working groups and deliver outcomes and reports from the groups.

Also evident is a subtle but growing presence from new players, including companies based in Asia and the Middle East keen to spread their wings and take a slice of the global market traditionally dominated by Western European, North American and Australian companies.

China has demonstrated an acceleration of environmental consultancy work with its rapidly growing environmental market. The Hong Kong Trade Development Council (HKTDC) Research (2020) noted:

> China's environmental industry has been developing in leaps and bounds in recent years. Investment in water conservancy, environment and public-facilities management amounted to RMB6.86 trillion in 2016 [$970 million], an increase of 23.3% year-on-year. Investment in the environmental sector is projected to exceed RMB15 trillion during the *13th Five-Year Plan* period (2016–2020), with the focus of

industrial development shifting from environmental pollution control to environmental quality improvement. Moreover, a large number of environmental projects. . .will be developed through public-private partnerships (PPP). . .The country's extensive economic development has resulted in serious environmental pollution and the concomitant economic cost and social problems have aroused great concerns from the government and the public. The demand for cost-effective solutions is therefore very keen on the mainland, forming a huge market for environmental protection service providers. . .The new Environmental Protection Law, which came into effect on 1 January 2015, aims to tackle environmental pollution issues. . .As of today, a number of leading enterprises are already well-positioned in China's environmental market. They include Thunip Corp Ltd, Zhonghang Yinyan, and Guangzhou Yueshou Environmental Holdings.

At the ENDS 2018 consultants' roundtable, key issues noted within the ENDS environmental consultancy market review 2018 were as follows:

- Consultancies are evolving to meet the demands of a world where projects are larger and more complex; these projects are changing the nature of environmental consultancies themselves.
- The supply chain is becoming ever more important; consultants are required to quickly bring in large numbers of skilled staff at relatively short notice and to keep them busy when the project ends or goes on hold. Some companies are focusing on their skills supply chain rather than looking to keep on growing as a company.
- The ability to quickly attract and hire the best specialists at short notice is becoming increasingly important; skilled and well-connected specialists can pick or choose the projects they work on and this may be a new model for the whole sector – a large pool of independent specialists who are headhunted on a project by project basis.
- Mega projects are draining the pool of skilled individuals in areas such as ecology, but on project completion, there is a glut until the next mega project begins.
- Retained staff are often being asked to widen skill sets in project and programme management and reporting rather than be field based alone.
- There is room for improvement in terms of using a company's global pool of expertise effectively.

It must be noted that there has been an element of Western 'over-focus' in terms of global studies of the environmental consultancy market. Future career opportunities in the sector are likely to be truly global, with companies headquartering worldwide and handling multiple projects using a mixture of in-house and contracted specialist staff.

2.3   Key Organisations and Employers

In a market with high levels of competition and active mergers and acquisitions, there are inevitably many surveys ranking the top companies within the wider sector, as well as the sub-sector specialisms.

Environment Analyst noted in its 2019 research report 'Global Environment Consulting Strategies & Market Assessment':

> The global market for environmental consulting (EC) services jumped 6.2% during 2017 to reach $34.1bn (£25.3bn). . .This is almost double the growth in 2016. . . Signs are that 2018 and 2019 will see similar order – if slightly more restrained – increases.
>
> Growth among the top players has been aided by recovering mining and commodities markets. . .as well as strong demand in climate resilience and storm recovery work in the key North American market. Also contributing to the result was a return to more buoyant conditions in Asia-Pacific spurred by several factors, including China's tightening environmental standards and enforcement.
>
> However, it is also notable that among the cohort of the leading international EC players analysed in depth in the study (the 'Global 22'), growth is being spearheaded by those in the lower half of the rankings rather than those in the top tier. . .the mid-size multidisciplinary firms and smaller environmental/sustainability specialists are more often than not outperforming their larger fully integrated technical and professional service provider rivals. . .many of these firms actively participating in mergers & acquisitions to boost growth – some on a truly transformational basis. The Global 22's dominance of what is still a fragmented market is evident from their combined performance in 2017, which saw aggregated EC revenues increase by 6.4% year-on-year to account for 43.6% share of the total global sector.
>
> Yet despite the continued consolidation, only three players – AECOM, Tetra Tech and Jacobs – hold an individual market share in excess of 5%. Meanwhile, Arcadis and ERM complete the top five.
>
> Mega-merger activity also saw the entry lower down the rankings of SNC-Lavalin to the Global 22, thanks to its acquisition of Atkins in 2017, which sees their combined global EC operations now in 19th place. . .The Global 22 together employ approximately 82,000 EC staff (FTEs) worldwide, servicing around 220,000 EC contracts per year, with an average contract value of $77k.

Box 2.2   The So-called Global 22
AECOM
Tetra Tech
Jacobs
Arcadis
ERM
Golder
RPS
Ramboll
Wood E&IS
WSP
Stantec
HDR

(Continued)

Box 2.2   (Continued)

GHD
Sweco
Antea Group
ICF
Royal HaskoningDHV
Mott MacDonald
SNC-Lavalin Group
Cardno
SLR
WorleyParsons

Engineering News-Record noted in its report '2018 Top 200 Environmental Firms' that the top firms in 2018 were as follows:

Box 2.3   Engineering News-Record Top Firms

AECOM
Jacobs
Clean Harbors, Inc.
Tetra Tech
Fluor Corp.
Bechtel Corp.
Stantec
Salini Impregilo
Wood PLC
Suez North America
HDR
Arcadis
ERM
Black & Veatch
Garney Holding Co.
CDM Smith
Golder Associates
Leidos, Inc.
Kiewit Corp.
The Walsh Group Ltd.

In terms of specific sectors, there are differing 'top players'. In Environment, Health and Safety (EHS) in 2019, the website consultancy.org reported that 'the independent research firm Verdantix had surveyed 400+ senior executives in the Americas, Europe, the Middle East and Asia Pacific working in the environment, health and safety (EHS) landscape, with eight out of ten respondents working for corporations with revenues greater than $1 billion'. They reported on which EHS consultancy had the strongest capabilities in the market. The top firms were as follows:

Box 2.4   Top EHS Firms
Dupont Sustainable Solutions (DSS)
AECOM
ERM
Jacobs
Ernst & Young
Lloyd's Register
Arcadis
Clean Harbors, Inc.
DEKRA Insight
Ramboll
GHD Environment
Stantec
Golder
WSP Parsons Brinckerhoff
Amec Foster Wheeler/Wood Group
Tetra Tech
Antea Group
Trinity Consultants
Huco Consulting
Langan

Case Studies – Multinational Engineering and Environmental Consultancies

AECOM

AECOM became an independent company in 1990 through the merger of five firms – the name AECOM stands for architecture, engineering, construction, operations and management. Archinekt News noted in 2010 that 'From that moment forth, the growth did not cease as more and more firms were acquired and more and more disciplines were incorporated. In 2007, AECOM became a publicly traded company, and today it is one of America's biggest firms with more than $6 billion in annual revenue'.

The company's key focus is infrastructure. They have partnered with many major global projects including the Rio 2016 Summer Olympics, Abu Dhabi's new airline hub, Midfield Terminal and Hong Kong Science Park. The company notes that they have had involvement in 'the world's longest cable-stayed bridge, record-breaking sports events, the largest greenfield port development mega project, life-sustaining disaster recovery programs, and the tallest tower in the Western Hemisphere'.

They have regional and local offices; regional offices include Los Angeles, London, Singapore and Brisbane.

Tetra Tech

Tetra Tech, founded in 1966, was a provider of consulting and engineering services. It has 20 000 staff across 450 offices globally, with its headquarters in California. Its diverse market areas include water management, asset management, AI and advanced data analytics, design and engineering, infrastructure and international development.

Its projects include Chicago Airports Environmental Support, Desert Sunlight Photovoltaic Project, and Hawaiian Marine National Monument Plan.

Jacobs

Jacobs Engineering was founded in 1947 and has 52 000 employees across more than 50 countries with its headquarters in Dallas, USA. Its diverse work areas include water, transportation, infrastructure, programme management and advanced technologies.

Its projects include Sustainable Water Initiative for Tomorrow (SWIFT) in the United States, Tuas Water Reclamation Plant & Integrated Waste Management Facility in Singapore and Aurangabad Industrial Smart City (AURIC) in Delhi–Mumbai, India.

Case Studies – Specialised Consultancies

Ecus

Ecus (Environment Analyst 2018a) was founded in 1986 in Sheffield and formed as a University 'spin-off'. Its areas of work include water management, ecology, EIA, development, and infrastructure. It has responded swiftly to market ebbs and flows by building or reducing its work in different areas such as construction, ecology, planning and renewables. It has a number of industry frameworks and has long-term contracts to deliver services for clients in the public and private sectors.

PJS Group

PJS Group is a rapidly expanding UK company, focusing on sustainable solutions to development and drainage challenges. The Group works in infrastructure design, geotechnical engineering, innovative and structural design and holistic land development. Its specialism is in finding innovative solutions and problem-solving. It has a young and dynamic team of specialists and actively recruits from universities in the United Kingdom.

Case Studies – Emerging Global Consultancies

Dar Group

Dar Group (2021) has over 19 000 employees operating from a total of 311 offices and 60 countries spanning most major cities in the Americas, Europe, Australasia, the Middle East, Africa and Asia. The company was founded in 1956 by four professors from the American University of Beirut.

Dar Group has a number of brands:

Dar Al-Handasah (Shair and Partners) – a multidisciplinary consultancy operating across the Middle East, Africa and Asia; the firm has deep expertise in planning, engineering, architecture, environment, project management and economics.

Dar established and shaped Dar Group into the global network of companies it is today. Brands in the group include Perkins and Will (Design, joined in 1986), T Y Lin (Global Infrastructure Engineering, joined in 1989), Currie & Brown (Project Management, joined in 2012), Penspen (Energy, joined in 1990), Integral Group (Global Building Engineering, joined in 2009), along with a number of specialised consultancy arms. In 2018, Dar Group acquired a strategic 25.9% equity stake in WorleyParsons.

Surbana Jurong

Surbana Jurong (n.d.) is one of the largest Asia-based urban and infrastructure consulting firms.

It is a Singaporean government-owned consultancy company formed in 2015 with the merger of Surbana International Consultants and Jurong International Holdings. It has acquired a number of other specialised consultancies, including Robert Bird Group (a specialist structural and civil and construction engineering consultancy), SMEC Holdings (consulting on major infrastructure projects), Sinosun Architects & Engineers in China and Singapore's KTP Consultants.

Headquartered in Singapore, the Surbana Jurong group has a global workforce of over 16 000 employees in more than 120 offices across over 40 countries in Asia, Australia, UK, the Middle East, Africa and the Americas. Surbana Jurong has a track record of close to 70 years and has built more than a million homes in Singapore, crafted master plans for more than 30 countries and developed over 100 industrial parks globally.

Guangzhou Yueshou Environmental Holdings

According to the Research in China website (2005–2011), Yueshou Environmental Holdings Limited is an investment holding company. The company, through its subsidiaries, operates in three segments: property development, environmental protection operations and forestry. The company's operations are located in Hong Kong and the People's Republic of China. In 2010, a wholly owned subsidiary of the company acquired Asiaone Forest Products Holdings Limited.

Thunip Corp., Ltd

Thunip Holdings was formed in 2002 from Tsinghua Unisplendour Environmental Protection Limited Liability Company, which was established in 1988 (edie Website 2020). Thunip's headquarters is in Beijing, China. The company has a focus on water and wastewater treatment technology, and the company has three arms: Thunip Water, Unisplendour Environmental Protection and Thunip Supply Chain. Key business activities are investment, construction and management of water and wastewater; engineering, procurement and construction (EPC) of environmental projects; and supply chain services of environmental protection products. Thunip owns many invention patents (e.g. its Rotary Fibre Disk Filtration System) and other advanced environmental protection treatment technologies. Thunip has completed over 4000 consulting projects, and more than 1000 engineering projects nationwide.

2.4 Careers

New legislation requires increased monitoring for emissions, waste and land redevelopment, and this has led to an increasing demand in employment opportunities. However, entry to the environmental consultancy profession is competitive. As noted earlier, consultancies operate globally and within multiple sectors, so research into a consultancy's projects, areas of work and strengths is always worthwhile before seeking employment opportunities.

Career progression in consultancy usually involves accepting a greater amount of office-based work. Effective marketing of the business and managing people, contracts and resources often takes over from the environmental work. Consultants operate in a very commercial environment, and senior staff may be required to help attract future clients for the business. However, larger consultancies are becoming more aware that they need technical specialists at a high level of management and to do so offers a pool of technical expertise, which can be used as a promotional/profile raising tool for the company. They can also operate with higher levels of government and regulators (often through technical panels, working groups and committees), and this can lead to sourcing new contracts that direct sales staff would not expose.

It would be sensible for someone seeking a role in the environment sector to engage with the larger firms in the sector initially, as they are well organised in terms of their recruitment process. Some have formalised graduate development schemes.

Many of the larger consultancies attend series of career fairs and visit universities internationally to seek candidates. They are also supportive of non-graduate recruitment in areas such as work experiences, internships and apprenticeships. There are regular changes in the areas where firms recruit, so it is worth keeping an eye on all the relevant companies. A success in the bid for a major project can lead to a targeted recruitment campaign in a specific area.

Table 2.1 Graduate development schemes and career information for 'environmental' consultancies.

Company	Scheme and company information	Links
AECOM	The company website has career paths and opportunities split into professional, craft/trade, military/veteran and early careers/graduates Register for information on The AECOM Talent Network www.aecom.avature.net/TalentCommunity	Global careers search: www.aecom.com/careers UK and Ireland www.aecom.com/uk-ireland-graduate-careers Middle East www.aecom.com/middle-east-graduate-careers Australia and New Zealand www.aecom.com/australia-new-zealand-graduate-careers Hong Kong www.aecom.com/hong-kong-graduate-careers Americas www.aecom.com/americas-graduate-careers

Table 2.1 (Continued)

Company	Scheme and company information	Links
Antea Group International	Within some office-specific Antea websites, there are website sections on careers, internships, campus recruiting and graduate opportunities	www.anteagroup.com/en/article/careers-0 Offices: www.anteagroup.com/en/article/our-offices Office-specific career information includes: India: www.anteagroup.co.in/en/article/careers-1 USA: www.us.anteagroup.com/en-us/careers
ARCADIS	The company has a global interactive opportunities map and a talent network where you can upload information and your CV	www.arcadis.com/en/global/careers www.arcadis.com/en/global/careers/graduates-interns
BSR	Innovative global consulting firm	www.bsr.org/en/sustainability-consulting www.bsr.org/en/careers
Bechtel	The company has an internship and new graduates programme, as well as apprenticeships and recruiting, for professionals, craft professionals and veterans	www.jobs.bechtel.com
Black & Veatch	The company has campus/university programmes, internships and the EDGE programme, which develops new recruits, as well as 'Workforce of the Future'	Career search: www.bv.com/careers www.bv.com/careers/campus www.bv.com/workforce-future
CDM Smith	The company has a Leadership Academy, 'Re-Boot Re-Entry' Programme in STEM, and programmes in STEM scholarship, Students and Graduates and Internships	www.cdmsmith.com/en/Careers www.cdmsmith.com/en/Careers/Students-and-Recent-Grads
Cardno	The company has graduate, internship and sponsorship programmes, as well as an Asia Pacific Graduate Development Program and an Americas Young Professionals Program (YP Program)	www.cardno.com/careers/ Careers search: www.cardno.com/careers/browse-careers/ Graduate roles and internships: www.cardno.com/careers/graduates-and-interns/
Clean Harbors, Inc.	The company website has a global job search and a Talent Network to upload information and your CV	www.careers.cleanharbors.com/en Job search: www.careers.cleanharbors.com/en/jobs
Dar Group	The company website has an 'open positions' section and a standard application form	www.dar.com/careers/ourculture www.dar.com/Careers/OpenPositions

(Continued)

Table 2.1 (Continued)

Company	Scheme and company information	Links
DEKRA Insight		www.dekra.com/en/home/ North America Career Centre: www.dekra.us/en/careers/
Dupont Sustainable Solutions (DSS)		www.consultdss.com/careers/ Global job search: www.consultdss.recruitee.com/
EcoVadis	Sustainable supply chain/rating provider with environmental consulting roles all over the world	www.ecovadis.com www.ecovadis.com/careers
Ecus	Ecus have internships and current vacancy listings on the careers page of their website	www.ecusltd.co.uk/ www.ecusltd.co.uk/about-ecus/careers/
Environ		www.environltd.co.uk/about-us/
Environmental Resources Management (ERM)	The company has a graduate entry programme and a staff partnership model, where ERM partners are the shareholding managers and leaders of the firm. There is a staff development programme and online ERM academy	www.erm.com/careers/ Graduate and entry level: www.erm.com/careers/graduate-and-entry-level/
Ernst & Young		www.ey.com/en_gl/careers
Fluor Corp	The company has their own 'Fluor University' offering accredited continuous learning courses	Global search: www.fluor.com/careers Graduate careers: www.fluor.com/careers/graduate/be-challenged-graduate
Garney Holding Co.	The company has a set of goals and a philosophy for staff development including GarneyU for online learning and an active community involvement programme	www.garney.com/careers/ www.garney.com/careers/current-openings/
GHD (inc. CRA)	The company has a Graduate Development Program and the 'GHD Business School', which assists with CPD and training	Global job search: www.ghd.com/global/careers www.ghd.com/global/careers/graduates/
Golder Associates	The company website has a global jobs search and a Talent Community	www.golder.com/careers/ www.golder.com/careers/graduates/
HDR	The company have internships and attend recruitment events, as well as have a monthly 'career starter' email service	www.hdrinc.com/home www.hdrinc.com/careers Job search: www.hdr.taleo.net/careersection
Huco Consulting		www.hucoinc.com/ www.hucoinc.com/about/careers/

Table 2.1 (Continued)

Company	Scheme and company information	Links
ICF	The company website has a careers section split into job type such as scientists, with a 'hot jobs' section: www.icf.com/careers/scientists	www.icf.com/company/about www.icf.com/careers
Jacobs	The company attends careers events, has employee network groups in areas celebrating inclusion and diversity, as well as an early careers programme that includes internships and apprenticeships	www.jacobs.com/ www.jacobs.com/careers Global job search: www.jacobs.taleo.net/careersection
Kiewit Corp.	The company has a Talent Community you can join by email, an internships programme and an entry-level jobs section on its website	www.kiewit.com/ www.kiewitcareers.kiewit.com/
Langan	The company has an internship programme, a training and development programme and mentoring available to all levels of staff	www.langan.com/ www.langan.com/careers Job search: www.langan.com/jobs/
Leidos, Inc.	The company offers over 400 internship and entry-level jobs a year. They are active in attending careers events, have a career path and leadership development programme	www.leidos.com/ www.careers.leidos.com/
Lloyd's Register	The company has a learning academy and a career development portal called Careerbox	www.lr.org/en-gb/ www.lr.org/en-gb/careers/ Job search: www.jobs.lr.org/search
Mott MacDonald	The company has an early careers programme covering ANZ Early Professionals and Interns, Hong Kong graduate positions, UK apprenticeships, UK and Ireland graduates and students and US graduates and interns. It also has a careers blog	www.mottmac.com/ www.mottmac.com/careers
PJS Group	The company tends to recruit direct from universities in the United Kingdom	www.pjsconsultingengineers.co.uk/contact/
Ramboll	The company has a mentoring scheme, student jobs and internships, as well as inviting collaboration for student projects and masters theses	www.ramboll.com/ www.ramboll.com/careers
Royal Haskoning	The company has a Young Royal HaskoningDHV platform to encourage networking, their career Global Positioning System (GPS), an international knowledge network and internships	www.royalhaskoningdhv.com/ www.royalhaskoningdhv.com/careers

(Continued)

Table 2.1 (Continued)

Company	Scheme and company information	Links
RPS	The company website has a job search and they recruit graduates, interns, trainees, apprentices, and consultants	www.rpsgroup.com/ www.rpsgroup.com/careers/
SLR	The company has regional websites and an international job search	www.slrconsulting.com/ www.slrconsulting.com/careers
SNC-Lavalin	The company website Early Careers section outlines internships, apprenticeships, work experience and graduate positions in different regions	www.snclavalin.com/en www.careers.snclavalin.com/
Stantec	The company has a students and graduates programme, internships (website has an international search) and comprehensive talent management and blended learning programmes for staff development	www.stantec.com/careers
Suez North America	The company has apprenticeships, volunteers for international experience and internships through its Emerging Workforce programme	www.suez-na.com/en-us www.suez-na.com/en-us/careers
Surbana Jurong	The company has internship programmes in various disciplines	www.surbanajurong.com/work-with-us/ www.jobstreet.com.sg/career/surbanajurong.htm
Sweco	The company website has links to career opportunities in all their offices	UK www.sweco.co.uk/career/ Job search: www.careers.sweco.co.uk/
Tetra Tech	The company have regular live Facebook events to discuss careers, a Leadership Academy, project management training programme and a professional development programme	www.tetratech.com/careers Job search: www.tetratech.referrals.selectminds.com/
Trinity Consultants	The company actively recruits from universities – the majority of new employees are hired this way in the United States. They have a Trinity Consultants Career Programme	www.trinityconsultants.com/ www.trinityconsultants.com/about/careers
WSP	The company has a professional growth network	www.wsp.com/ www.wsp.com/careers www.wsp.com/careers/early-career
The Walsh Group Ltd	The company has an entry-level programme that includes internships	www.walshgroup.com/ www.walshgroup.com/careers.html www.entrylevel.walshgroup.jobs/

Table 2.1 (Continued)

Company	Scheme and company information	Links
Webuild (formerly Salini Impregilo)	The company operates internships and a programme for students and new graduates	www.webuildgroup.com/en/careers Job search: www.jobs.salini-impregilo.com/
Wood Group	The company has Early Years recruitment and a Developing Professionals Network	www.woodplc.com www.woodplc.com/careers
Worley	The company has a 1000-strong early careers network	www.worley.com/ www.worley.com/careers

Source: Adapted from www.aecom.com/careers UK and Ireland www.aecom.com/uk-ireland-graduate-careers Middle East www.aecom.com/middle-east-graduate-careers Australia and New Zealand www.aecom.com/australia-new-zealand-graduate-careers Hong Kong www.aecom.com/hong-kong-graduate-careers Americas www.aecom.com/americas-graduate-careers. © John Wiley & Sons.

2.5 Job Titles in the Sector

Consultancies have become multidisciplinary organisations, covering many areas of work from civil engineering to ecological surveys to environmental policy and strategy development. There are too many titles to list, but they can include:

Consultant, Senior Consultant, Associate, Director, Planning Consultant, Engineer, Modeller, Policy Specialist, Civil Engineer, Water Management Specialist, GIS Manager, Sustainability Manager.

2.6 Educational Requirements

In its article 'How to become an environmental consultant', the website CareerExplorer suggests the following educational requirements:

- Environmental consultants will typically have an undergraduate degree and sometimes even a master's degree in environmental engineering, environmental science, environmental studies, or some other science discipline
- Employers will often give preference to candidates who have earned a master's degree

2.7 Personal Attributes and Skill Sets

There are many roles in the sector, but the following skills and experience are helpful:

- Some form of experience, either through volunteering, a work experience or internship.
- Data analysis and reporting.
- Communication and presentation skills, including presentations to committees and the public.

- Teamwork and leadership, including co-ordination of teams.
- IT skills, including GIS, mapping and modelling in some roles.
- Project management and options appraisal.
- Finance and budgeting administration.
- Specific sector knowledge, relevant to your preferred roles.
- Ability to communicate scientific data and findings to other audiences.

2.8   Career Paths and Case Studies

◀ Top Tips

Phil Le Gouais – Practice Leader – Environmental Impact Assessment (Mott MacDonald)

1) Be open minded about what you do – If you have too fixed an idea of what you specifically want to do – for example at the start of my career I really wanted to work in Africa – but the opportunities that came up for me seemed to be everywhere but Africa. I still took them – holding out for the perfect position means you are missing lots of other rewarding experiences.
2) Do not just know what you are doing – know why you are doing it. You can make yourself much more useful if you understand the context of why your employer or your client is asking you to do it. In my view that is the key to 'adding value'.
3) Know why you are in the game – I have three values that I follow about my career, and I measure all of my career moves against those. I do not want to spend all my time at work (or on my way to work), I want an interesting life and I want the opportunity to change the world. If it passes those three tests, it is worth seriously considering. For some people, it is money; for some people, it is problem-solving; for some people, it is status. They are all fine, just be honest with yourself or you will end up in a dead end.
4) Manage your career brand – if you are deeply technical – present yourself as a guru – if you are deeply commercial, present yourself as an entrepreneur, if you are creative, be a visionary. It is useful to be an all-rounder – but it is a bit dull.
5) Be prepared to quit your job tomorrow – I really like my job, and I've been with my company for 15 years, but if something better came along, I'd still test it against my three principles (above). I can also think of three times in my career my job stopped meeting my expectations and each time I've quit and moved on (admittedly twice I've come back to Mott MacDonald) – but at least I found out!

Source: Phil Le Gouais – Practice Leader – Environmental Impact Assessment (Mott MacDonald). www.linkedin.com/in/fernandopenharebelo. © John Wiley & Sons.

◀ Top Tips

Fernando Rebelo, Project Manager and Environmental Consultant at DEKONTA a.s., Czech Republic

1) The first one is for undergraduates. Study, live your social life and make good connections with older students while at university. If there is something that I wish I had

done differently when I was a student at university is the fact that I have not thought at the time that building future connections was so important. It is very important. These are the people that will enter the job market before you and could potentially recommend you for positions within their companies.

2) The second tip is pretty much for everybody in my area. Usually positions in environmental consultancy tend to be very specific. Sometimes it feels that no vacancies match your experience and even that there are no positions at all. But one thing is that the majority of the jobs are not advertised. This means that the advertisements in a magazine, in a website or on LinkedIn are only the final steps of a recruitment process that already started weeks or even months ago with a decision-maker asking their colleagues if they could recommend somebody for a new position. So, my tip is to try to present yourself to companies, attend open days, speak with experts that give presentations in congresses or workshops and do not be afraid to just call the relevant department's director if there is a vacancy. This is actually how I got all my student jobs (by knocking on doors and handing in CVs) and my current graduate job (I called the company's director and asked if there was a position opened and he called me for an informal chat).

3) The next tip is for recent graduates. You will learn much more in the first month at your job and at your placement year than in all your university lessons combined. Yes, it is the uncomfortable truth. So, when you are in your new graduate job, try to expose yourself to as much information as possible. Try to help a colleague from a different department with his project or try to offer some of your time to a colleague doing a different task. You will learn a lot.

4) Never be afraid to receive a no for an answer. Do not be shy to ask for a favour. Imagine the 'no' as the default answer and everything other than that as a bonus. This means that, unless you are extremely well connected, chances are that your job, promotion or opportunity to work in a new project will not fall on your lap. You might have to ask for it. So, if you feel that you are ready and deserve the opportunity, just ask. You might have a pleasant surprise.

5) Finally, it might seem a cliché, try to find a job that you really like. Not all in life is money driven. Yes, we all have to pay our bills, but we also need to find a motivation to wake up early every morning whether it is sunny, raining, snowing or even if you did not sleep the previous night because both your daughters were awake the whole night and did not let anybody sleep. You still need to find the motivation to get out of bed and go to work. So, finding a position with a salary that enables you to live comfortably, where the team is a pleasure to work with and that allows you to have a good work–life balance is extremely important.

Personal Profile

Born in Brazil but raised in Portugal, Fernando Rebelo attended university in Portugal, Greece and England and currently lives in the Czech Republic where he works for DEKONTA and is responsible for the implementation of Development Projects for clients like UNDP, UNIDO, EBRD. He occasionally offers his experiences for other companies looking for independent short-term consultants.

His main areas of expertise are:

- Persistent Organic Pollutants and Obsolete Pesticides
- Transboundary Movement of Waste

- Basel, Rotterdam and Stockholm Conventions
- ADR and IMDG Regulations
- Capacity Building and Institutional Strengthening
- International Development.

Always willing to explore, he is the happiest doing what he does best, which is to support developing countries in finding sustainable solutions for their environmental problems.

Personal Profile

Ogega, Bosibori. M, Environmental and G.I.S Consultant, Ecolife Consulting Ltd, Kenya

My current role
- Environmental information communication for stakeholder forums
- ESIA/SEA/EIA/EA report processing and synthesis
- Data collection and analysis
- Environmental database management
- Geospatial data analysis

Required skills
- Educational background in Geography and Environmental Studies
- Registration as an expert by the National Environmental Management Authority

How I ended up in this role
- Studied geography at university undergraduate level because of a passion I have had since I was a child. I realized I loved nature at the age of 11. I have mostly shared this interest with my father who I accompanied for most trips to the wild while he was a secondary school geography teacher. After joining Ecolife Consulting Ltd, an environmental firm, I had enough funds to start and finish my MA in Environmental Planning and Management after which I rose to be a senior resident consultant.

Tips and advice
- This is especially for people aspiring to join the field in East African countries: the environmental field is still relatively new in this region and government institutions are still not very responsive towards environmental matters, especially in towns that are not capital cities. Therefore, it is not a highly valued profession like most, ensure you enjoy doing it because it will not make you wealthy very fast. It is probably the least understood career in Kenya. Do it for the love of the Earth, the goodness of Nature and for the future of Mankind!

Source: Ogega, Bosibori. M, Environmental and G.I.S Consultant, Ecolife Consulting Ltd, Kenya. © John Wiley & Sons.

◄ **Top Tips**

Peter Jones, Ecolateral, UK

1) Opportunities for adaptive development
2) Ability to integrate sociological, technological and economic trade-offs
3) Cross-sectoral experience
4) Ability to analyse data and facts
5) Awareness of recent environmental science findings

Source: Peter Jones, Ecolateral, UK. © John Wiley & Sons.

Personal Profile

Monica Elizabeth Salirwe, Senior Environmental Consultant/Head, Sustainability and Policy Unit, Uganda

I ended up in the environmental career majorly because of my education background (nature conservation). I did an internship for a campaign called Save the Planet and thereafter got a job through a recommendation from one of my university lecturers – so, my current job has shaped my career path. My job typically entails a lot of technical work and the occasional administrative-related work. Basically, I do a lot of technical report writing on several issues such as Environmental Assessments and Audits, Environmental and Social Management Systems, Cleaner Production and Resource Efficiency aspects, Waste Management aspects, Sustainability issues for a variety of sectors such as oil and gas, agriculture, hydro and solar power, among others. Several skills are needed for a career in the environmental sector, but I will focus on the soft skills that I have found to be very crucial, and yet not overly emphasised, and these include:

- Critical thinking ability
- Ability to communicate technical environmental issues to a layman (someone who does not have a shared background in environmental issues)
- Innovation and alternative thinking (thinking beyond the norm)
- The ability to work independently and make independent decisions.

I would advise someone interested in an environmental career to be very keen on internships and volunteer work as they provide an avenue for exposure into several environmental sectors, learning practical elements, and through that, one can easily identify what they are passionate about and where best they would fit in and contribute most.

Source: Monica Elizabeth Salirwe, Senior Environmental Consultant/Head, Sustainability and Policy Unit, Uganda. © John Wiley & Sons.

Company Profile

Azura, Canada

Azura Associates International Inc. was established by David Ellis in 2012 to provide specialist consulting services to industrial wastewater treatment facilities, anaerobic waste-to-energy facilities and innovative cleantech businesses. Backed by decades of experience, Azura provides proprietary testing, research and development and engineering services. Unlike many large consultancies that focus on large capital construction projects, Azura focuses on improving the performance of existing systems and infrastructure.

Source: Azura, Canada. www.azuraassociates.com. How to Build your Environmental Consulting Career: Advice for Young Professionals. © ECO Canada.

Personal Profile

David Ellis, Azura Associates International Ltd, Canada

David Ellis, a chemical engineering consultant with more than 25 years' experience in the wastewater industry, is a prime example of someone who values the importance of self-investment for a career in consulting. David's passion for cleantech and wastewater, as tools for improving lives and protecting the environment – combined with a knack for personalized communication – led him to launch Azura Associates International Inc. in Canada.

According to David, the secret to a great consulting career is a combination of technical and interpersonal communication skills, being a translator for customers who are inexperienced in the technical issues.

David actively mentors young professionals in the local water environment association. He notes that new employees, graduates and students must realize that they are *active managers of their own career.* Each person is responsible for building their skillset and reputation to enhance career possibilities.

His advice to young professionals – build your career on these *three pillars:*

- *Professional skills and experience*
- *Professional profile*
- *Professional network.*

Your professional skills and experience are the foundations of a successful career. Your diploma or degree represents the basic knowledge required to enter a profession. After that, you may need on-the-job training and additional certifications depending on your position. But it is important that you do not stop there. Changes will occur in technology, regulations, and more, and you should remain in a constant state of learning.

Take advantage of any auxiliary training offered by your employer. Continue on your own by taking specialized courses and seminars. Join your professional associations and attend educational activities. Read newsletters, journals and magazines. Take a self-study course. If your employer will not pay the fees, do so yourself. Remember, it is your future and you are the one in control of your career path.

The most common mistake is to develop depth at the expense of breadth. You may feel ready for a promotion and salary increase after five years of experience, but you may have fallen into a common trap: instead of having five years of experience, you've simply got one year of experience five times.

Ensure you are being stretched with your assignments and you are proactive with your supervisor. It is all right to do a similar project once to learn from your mistakes, but push for some new challenges. Managing your experiences with your employer is important so that you are ready for advancement.

Your professional profile relates to how people you have never met perceive you. In today's internet-intensive culture, your professional profile is extremely important. Employers and clients often use this information when making decisions about hiring, promotions and more.

Part of maintaining a positive profile is keeping your online presence tidy and professional.

Having a presence on a professional network, such as LinkedIn, and participating in discussions related to your field can be helpful.

Again, joining your technical association and participating in online articles or discussions may boost your profile. Contribute a newsletter article or do a guest blog on one of the sites to get your name out there in a positive way.

Also, by volunteering with your association to help prepare for a group activity, you become known as a person who is doing good things for the profession and the community.

Your professional network refers to the people you have met. Your professional network encompasses people you work with every day, those you may only see on occasion and those you've only met once. Whatever the relationship, getting involved and raising your profile in that community will go far in its effect on your profession. With each encounter, you will leave an impression that may have a lasting effect on your career.

Being active in the community and in your professional associations is a powerful way to widen your network. For instance, by volunteering to assist at a technical conference, you will meet industry leaders. By acting as an association officer or newsletter editor, you will show your reliability and concern for the profession and you will both get to know others and become known.

Good communication and social skills are important to building your network. You may want to take some courses on public speaking. The ability to address a crowd and explain your ideas clearly will build your confidence. Write a technical paper. Teach a class. Eventually you may become an industry leader.

Together, these three pillars will build credibility and trust in the work you do. In the modern world, the days of forever employment are long passed. So continual investment in yourself is the best investment you will ever make.

For technical careers with Azura, an engineering degree is a must; a general science or environmental studies degree is of little value. We work in a highly regulated environment, so the ability to become a licensed professional who can certify technical reports is essential. Many reports are submitted to government regulators. The regulations stipulate which professionals are allowed to certify the submittals, and in many cases, in my area, only licensed Professional Engineers are permitted to make that certification.

When hiring, I look for industrial and factory experience, not consulting experience. I want people who have the ability to relate to Azura's industrial clients, who understand what it is like on a production line or what it is like in a manufacturing facility that runs 24/365. The more smelly, loud, chaotic and dirty the factory, the better the experience. After joining Azura, people have little opportunity to gain that immersive exposure to a factory environment. We often visit factories for only two or three days. During that short time, we need to be instantly comfortable in that environment, and we need to establish excellent rapport with all plant staff, from the most junior production worker to the most senior executive.

Source: David Ellis, Azura Associates International Ltd, Canada. © John Wiley & Sons.

◀ **Top Tips**

Marita (Ariel) Oosthuizen, Principal Scientist, GA Environment (Pty) Ltd, South Africa

The Environment is a very wide field and can be approached from any undergraduate area. Bearing this in mind, there are a few tips that I can give.

1) Know your aptitude, strengths and interests before you make a study choice. In other words, decide if you want to come into the environmental field from an engineering, humanities, biophysical, economic or pure sciences angle. Then do a first degree (including environmental subjects) from that angle. This will allow you to specialise at a later stage.

2) If you are not sure or you feel strongly that you want to be in a pure environmental/ sustainability field courses that focus solely on environmental subjects have developed and then you should invest in those.

3) Finances and life circumstances allowing, do not stop before you have at least a master's degree.

4) Except if you are 100% sure what you want to specialise in, do not narrow your focus too early. Work in an environment where you can try out a number of avenues (e.g. EIAs and sustainability work) and work as a generalist until you have found your 'sweet spot' where you can make a difference, have career opportunities, feel fulfilled and grow as an individual.

5) By all means, start as a generalist – I really recommend that – but do not remain one. Everyone who has ever succeeded in life figured out the one thing s/he does better than anyone else and focused on that until s/he is THE expert in the field.

6) Do remember to keep studying after you started working and ensure that you grow in terms of your management, leadership and governance abilities as well.

Personal Profile

My passion is with psychology and geography. I studied both pre-grad and continued with environmental management in my Honours and Masters year. My first job opportunity was as an EIA consultant where I dealt with the entire EIA process (both public participation and the technical work). I excelled at public participation and soon started working at a niche company that specialised in public participation.

However, eventually, I felt as if I worked myself into a 'public participation corner' and could not do the technical work anymore. I managed to work in the Environmental Management Systems field and environmental awareness training fields for a while. I am again working as an EIA generalist (doing both the public participation and technical work). I have developed an interest in social impact assessments now, and I am slowly developing in that direction as I see it as a new direction for me.

Currently, I am developing environmental awareness training programmes, doing EIAs and acting as a specialist in terms of developing public participation processes and facilitation. I am also doing a bit of management, and I am developing skills in terms of social impact assessments. I find this mix more stimulating – although I am a staunch believer in focusing on one or two (related) things that you excel in.

Source: Marita (Ariel) Oosthuizen, Principal Scientist, GA Environment (Pty) Ltd, South Africa. © John Wiley & Sons.

Top Tips

Tony Barbour, Head of Tony Barbour Environmental Consulting and Research, South Africa

1) As an environmentalist I think it is important to have a bit of activist in you, so you need to be passionate about environmental matters and aware of key environmental issues. At the end of the day, you should be looking after the interests of the environment, not just your client.
2) Stay in touch and connected, linked to the above, one needs to be aware of what is happening in the environmental field and what the key issues are.
3) Understand the linkages and relationships between environmental, social and economic issues and challenges and what the potential trade-offs are, and the potential implications (short, medium and long term) of those trade-offs are.
4) Understand and acknowledge that different people and societies have different world views that may not necessarily be the same as yours.
5) Appreciate the role that technology can play in addressing environmental challenges, but at the same time understand the uniqueness of environmental ecosystems and services, and the fact that some impacts may be irreversible and cannot be undone by technology.

Personal Profile

I did a BSc in Economics and Geology, and Honours in Economics, but then I decided to do a master's degree in Environmental Science. When I was writing up my thesis, I managed to get a job as a junior environmental consultant with a local engineering company that had an environmental section (one of the few in South Africa in the early 1990s). Spent five years there, then I got approached by another engineering company to start up an environment section in the Cape Town office, so started and managed that section for 4½ years.

I got a bit tired of consulting and was lecturing part time at the University of Cape Town, which had an environmental research unit attached to the department of Environmental and Geographical Science. They offered me a post as a senior researcher, where I spent the next 4½ years. Great place to be but took a large cut in salary when I joined from the private sector. So after 4½ years, I decided to give it a go working for myself and 14 years later seems to be working out ok.

My main focus is on social impact assessments and review work. My company is Tony Barbour Environmental Consulting and Research.

Source: Tony Barbour, Tony Barbour Environmental Consulting and Research. www.tonybarbour.co.za. © John Wiley & Sons.

2.9 External Resources

UK

Prospects UK Job Profile:
www.prospects.ac.uk/job-profiles/environmental-consultant

National Careers Service, UK:
www.nationalcareers.service.gov.uk/job-profiles/environmental-consultant

Planit, UK:
www.planitplus.net/JobProfiles/View/4/14

Earth How (2020): 'The Good, the Bad and the Ugly for Environmental Consulting Careers'
www.earthhow.com/environmental-consulting-career/

The Guardian Environment Jobs:
www.jobs.theguardian.com/jobs/environment/

Europe

Morning Future, Italy
www.morningfuture.com/en/article/2019/04/08/environmental-consultant-green-economy-work/590/

Environmental-Expert.com 'Environmental Consulting Companies in Europe'
www.environmental-expert.com/companies/keyword-environmental-consulting-8942/location-europe

Asia

Environmental-Expert.com 'Environmental Consulting Companies in Asia'
www.environmental-expert.com/companies/keyword-environmental-consulting-8942/location-asia-middle-east

Nallathiga, R. (2014). Market for environmental products and services: a review of Asia Pacific Countries. *TIDEE (Teri Information Digest on Energy and Environment)* 13(2): 163–176. https://www.researchgate.net/publication/273895000_Market_for_Environmental_Products_and_Services_A_review_of_Asia_Pacific_Countries

Africa

Acer Africa
www.acerafrica.co.za

Consult 5
www.consult5.com/Africa_Service.html

Africa Work
www.africawork.com

North America

Academic Invest 'How to Become an Environmental Consultant'
www.academicinvest.com/science-careers/environmental-science-careers/how-to-become-an-environmental-consultant

Career Explorer
www.careerexplorer.com/careers/environmental-consultant

Day in the Life of Woo Lim, Environmental Consultant
www.youtube.com/watch?v=HGxBERhVaFQ

South America

Environment Analyst – Latin America News
www.environment-analyst.com/global/latin-america

Environmental Expert – environmental consulting services in Latin America
www.environmental-expert.com/services/keyword-environmental-consulting-8942/location-latin-america

Oceania

Environmental Consultants Association 'Mega Trends in Environmental Consulting in Western Australia 2018 Workshop'
www.eca.org.au/news-activities/eca-past-events/mega-trends-in-environmental-consulting-in-western-australia-2018-workshop

Environment Analyst (2016) 'Country Insight: the rise, fall & recovery of Australia's environmental services sector'
www.environment-analyst.com/global/47038/country-insight-the-rise-fall-recovery-of-australias-environmental-services-sector

Environmental Expert – environmental consulting Companies in New Zealand
www.environmental-expert.com/companies/keyword-environmental-consulting-8942/location-new-zealand

References

Academic Invest Website (n.d.). How to become an environmental consultant. www.academicinvest.com/science-careers/environmental-science-careers/how-to-become-an-environmental-consultant (accessed 13 April 2021).

Allaboutcareers.com (2004–2020). Environmental consultant • job description, salary & benefits. www.allaboutcareers.com/careers/job-profile/environmental-consultant (accessed 13 April 2021).

Archinekt News (2010). Who the hell is AECOM? www.archinect.com/news/article/99869/who-the-hell-is-aecom (accessed 13 April 2021).

Business Insider (2017). 9 of the world's largest megaprojects that are under construction. www.businessinsider.com/the-worlds-largest-megaprojects-2017-1?r=US&IR=T (accessed 13 April 2021).

Career Explorer (2021). How to become an environmental consultant. www.careerexplorer.com/careers/environmental-consultant/how-to-become (accessed 13 April 2021).

Consultancy.org (2019). The top consulting firms for environment, health and safety (EHS). www.consultancy.org/news/157/the-top-consulting-firms-for-environment-health-and-safety-ehs (accessed 13 April 2021).

Dar Group History (2021). Dar Al-Handasah's journey: sharing in owning a sustainable future. www.dar.com/about/history (accessed 13 April 2021).

edie Website (2020). Thunip Holdings Co. Ltd. www.edie.net/59138/d/Thunip-Holdings-Co-Ltd (accessed 13 April 2021).

Engineering News-Record (2018). ENR 2018 top 200 environmental firms 1–100. www.enr.com/toplists/2018-Top-200-Environmental-Firms-1 (accessed 13 April 2021).

Environment Analyst (2018a). SME insight: Ecus – under new management. https://environment-analyst.com/global/63729/sme-insight-ecus-under-new-management?q=Ecus+under+new+management (accessed 13 April 2021).

Environment Analyst (2018b). Strengthening recovery for global environmental consulting market. https://environment-analyst.com/global/62991/strengthening-recovery-for-global-environmental-consulting-market?q=strengthening+recovery (accessed 13 April 2021).

Environment Analyst (2019). Global Environment Consulting Strategies & Market Assessment report 'Global environmental consulting market surges'. www.environment-analyst.com/global/73196/global-environmental-consulting-market-surges (accessed 13 April 2021).

Environment Analyst Business Summit (2019). CEO question time, John Chubb RPS. https://environment-analyst.com/global/77869/ceo-question-time-john-chubb-rps?q=Chubb (accessed 13 April 2021).

Ernst & Young Report (2014). Spotlight on oil and gas megaprojects. http://globalsustain.org/files/EY-spotlight-on-oil-and-gas-megaprojects.pdf (accessed 13 April 2021).

Flyvbjerg, B. (2014). What you should know about megaprojects and why: an overview. *Project Management Journal* 45 (2): 6–19.

HKTDC Research (2020). China's environmental market. https://research.hktdc.com/en/article/MzA4NzY1NDAz (accessed 13 April 2021).

McKinsey & Company (2015). Megaprojects: the good, the bad, and the better. www.mckinsey.com/industries/capital-projects-and-infrastructure/our-insights/megaprojects-the-good-the-bad-and-the-better# (accessed 13 April 2021).

Pinsent Mason (n.d.). Joint ventures – delivering infrastructure projects. https://www.pinsentmasons.com/thinking/special-reports/joint-ventures-global-infrastructure-projects (accessed 13 April 2021).

Research in China Website (2005–2011). Yueshou Environmental Holdings Limited. www.researchinchina.com/Htmls/Company/3649.html (accessed 13 April 2021).

Salvidge, R. ENDS Report (2018). Consultancy market review 2018: the changing face of environmental consultancy. www.endsreport.com/reports/ecmr2018 (accessed 13 April 2021).

SNC Lavalin Press Release (2017). SNC-Lavalin completes transformative acquisition of WS Atkins. www.snclavalin.com/en/media/press-releases/2017/03-07-2017 (accessed 13 April 2021).

Stantec Press Release (2016). Stantec to acquire MWH, a global professional services firm with leading expertise in water resources infrastructure. www.stantec.com/en/about-us/news/2016/stantec-to-acquire (accessed 13 April 2021).

Surbana Jurong Website (2019). www.surbanajurong.com/ (accessed 13 April 2021).

Wood Group Press Release (2017). Wood Group completes acquisition of Amec Foster Wheeler. www.woodplc.com/news/2017/wood-group-completes-acquisition-of-amec-foster-wheeler2 (accessed 13 April 2021).

3

Integrated Water Resources Management

3.1 Sector Outline

Growing populations and increasing water requirements are a certainty, while more and more water will be required for environmental concerns such as aquatic life, wildlife, recreation, scenic values and riparian habitats. Thus, increased competition for water can be expected and water resources management should be flexible so as to be able to cope with changes in availability and demands for water. This will require intensive management and international cooperation, where all significant factors are considered in the decision making process, considering not only supply management, but also demand management (water conservation, transfer of water to uses with higher economic returns, etc.), water quality management, recycling and reuse of water, economics, public involvement, public health, environmental and ecological aspects, socio-cultural aspects, water pollution control, sustainability (Bouwer 2000).

It is challenging to set boundaries for the water management sector. EU legislation the Water Framework Directive promotes a holistic approach to water management through river basin management, as do measures in other countries in watershed management. These approaches clarify the link between water management and land management and introduce and resource cross sectoral and multidisciplinary approaches to water management.

This chapter is a 'mega chapter' because of the growing integration of functions and organisational departments and the need to manage issues like catchment management, climate change impacts, flood risk management, extremes of weather, population change and accelerating urbanisation. Within the sector there is a myriad of roles and functions.

Even though the generic term used for the people who study water is 'hydrologist', people working in the sector include many roles such as hydrologists, water scientists, water quality specialists, hydrogeologists, environmental scientists, earth systems scientists and water resources planners.

'Hydrologists' work to solve water-related problems, such as:

- Pollution carried by water – including oceans, rivers, streams, rain, snow and ice – and devise methods to clean and control it
- Weather-related problems, such as flood forecasting and rainfall-runoff control

Global Environmental Careers: The Worldwide Green Jobs Resource, First Edition. Justin Taberham.
© 2022 John Wiley & Sons Ltd. Published 2022 by John Wiley & Sons Ltd.

- Drought management, acid rain and global warming
- Global water resources management so that all water uses – municipal (home and city), industrial and agricultural – are achieved efficiently while protecting the environment

For the purposes of this chapter, the following disciplines will be covered:

- Integrated Water Resources Management (IWRM) and integration within the sector
- Water Resources – including Hydrology
- Water Supply and Drinking Water Quality
- Wastewater Management
- Flood and Coastal Erosion Management
- Catchment, River Basin and Watershed Management

Integrated Water Resources Management (IWRM) and Integration Within the Sector

As noted, there is a more integrated approach being taken globally in water management. The Integrated Water Resources Management (IWRM) approach integrates knowledge and resources from a wide range of disciplines – engineering, geology, geography, hydrology, forestry, soil science, fisheries, biochemistry, law, advocacy, policy, education, operations, outreach, economics, tourism – and channels these resources and information to develop long-term sustainable economic, regulatory and conservation measures.

Water Management in the twenty-first century is a work in progress. Until recently, water resource management employed a fragmented systems framework. This created a decentralised management style, composed of differing groups, often operating independently and at odds with each other, with a focus on a short-term, one-at-a-time problem-solving approach.

However, due to growing worldwide concerns regarding water scarcity, access and quality, a universal consensus has emerged that water policy must take an integrated, long-term approach to solving these problems as quickly as possible before irreversible damage occurs to vulnerable populations and the global environment. This change in approach has led to the development of IWRM.

The Global Water Partnership (n.d.) has noted:

> Integrated Water Resources Management (IWRM) is a process which promotes the coordinated development and management of water, land and related resources in order to maximise economic and social welfare in an equitable manner without compromising the sustainability of vital ecosystems and the environment.

Conflicts of interest between environmental, social and economic needs should be addressed through moves towards sustainable development. A main focus for the water quality and wastewater industry is how to better conserve the existing water resource and find more efficient ways to make it safe for the population or to return to natural water sources.

In the EU, The Water Framework Directive (WFD) seeks to integrate assessment of water quantity and quality together with the ecological health of water bodies. This has led to increased use of the concept of water security, which seeks to integrate water availability under the extremes of floods and drought with ecological status.

However, there is the argument that the active promotion of IWRM by global organisations and funding bodies such as the World Bank and the African Development Bank has been a principle which in practice is very hard to implement, as highlighted in the working paper 'Flows and Practices: Integrated Water Resources Management (IWRM) in African Contexts' by Lyla Mehta and Synne Movik (2014):

> The promotion by these global players has led to a quasi-global industry around IWRM manifesting itself in various forms such as Master's degrees and short courses, annual symposia such as WaterNet in southern Africa, IWRM toolkits and manuals as well as major water reform programmes and the rewriting of national policies drawing on IWRM principles in a range of countries in the south. It has rapidly moved from the global north and temperate regions to many developing countries in Africa, Asia and Latin America. However, a growing body of research has highlighted that the experiences of IWRM in many developing country contexts have been mixed. In the African context, there is emerging evidence that it has not produced the anticipated socio-economic, political and ecological outcomes due to the uncertainty and complexity of river basins and the plural, overlapping and competing formal and informal legal and customary systems.

Water Resources – Including Hydrology

'Water resources' broadly relates to the assessment of the quantity, quality and variability of groundwater and surface waters. Addressing global impacts on water resources depends on the ability of professionals to make knowledgeable and sustainable management decisions. Water resources planning and management practices in the twenty-first century are increasingly viewed from a multi-disciplinary approach.

Water resource specialists provide project management on issues or assigned initiatives relating to the water resources of a city or municipality. They coordinate with other local jurisdictions regarding the water resource management program of the city or municipality, which encompasses issues such as surface water runoff, drinking water issues and conservation issues.

Water resources issues include local drinking water, large scale catchment management for water supply, irrigation water needs, hydropower schemes, flood and erosion control and protection of aquatic systems for wildlife.

Water Supply and Drinking Water Quality

The WHO (n.d.a) note:

> In 2017, 71% of the global population (5.3 billion people) used a safely managed drinking-water service – that is, one located on premises, available when needed and free from contamination. 90% of the global population (6.8 billion people) used at least a basic service. A basic service is an improved drinking-water source within a

round trip of 30 minutes to collect water. 785 million people lack even a basic drinking-water service, including 144 million people who are dependent on surface water. Globally, at least 2 billion people use a drinking water source contaminated with faeces.

Water suppliers abstract water from the environment, treat it to the required standard and distribute it to household and non-household customers. Traditionally this sector was bluntly called 'Clean Water'.

Wastewater Management

According to UN Water (n.d.) 'Due to population growth, accelerated urbanization and economic development, the quantity of wastewater generated and its overall pollution load are increasing globally.'
The World Water Assessment Programme (UNESCO WWAP n.d.) noted:

> Globally, it is likely that over 80% of wastewater is released to the environment without adequate treatment. . .The opportunities from exploiting wastewater as a resource are enormous. Safely managed wastewater is an affordable and sustainable source of water, energy, nutrients and other recoverable materials.

Traditionally this sector was called 'Dirty Water'.

Flood and Coastal Erosion Management

The WHO (n.d.b) highlighted:

> Floods are the most frequent type of natural disaster and occur when an overflow of water submerges land that is usually dry. Floods are often caused by heavy rainfall, rapid snowmelt or a storm surge from a tropical cyclone or tsunami in coastal areas. Between 80 and 90% of all documented disasters from natural hazards, during the past 10 years, have resulted from floods, droughts, tropical cyclones, heat waves and severe storms. Floods are also increasing in frequency and intensity, and the frequency and intensity of extreme precipitation is expected to continue to increase due to climate change.

In the World Bank Report 'Cities and Flooding' (2012) it was noted:

> In the past twenty years in particular, the number of reported flood events has been increasing significantly. The numbers of people affected by floods and financial, economic and insured damages have all increased too. In 2010 alone, 178 million people were affected by floods. The total losses in exceptional years such as 1998 and 2010 exceeded US$40 billion.
> Floods affect urban settlements of all types, from small villages and mid-sized market towns and service centres, for example along the Indus River, to the major

cities, megacities and metropolitan areas like Sendai, Brisbane, New York, Karachi and Bangkok, all of which were struck by recent floods.

With a solid understanding of the causes and impacts of urban flooding, an appreciation of the likely future flood probability and of the uncertainties surrounding it, and knowledge of both the potentials and the limitations of various flood risk management approaches, policy makers can adopt an integrated approach to flood risk management.

This integrated approach will direct future trends in careers in the water management and flood risk management sectors.

Catchment, River Basin and Watershed Management

The Ecosystems Knowledge Network (n.d.) states:

> The 'catchment-based approach' and 'integrated river basin management' are the terms often used to refer to the management of land and surface water as a *system*. The ecosystem approach highlights the importance of this management style, because of its emphasis on actions that reflect how nature works. It also encourages actions that result in people different interests in land and water working together in partnership to identify issues, agree what should be done and put this into practice.

In this way, catchment management is an analogy of Integrated Water Resources Management.

3.2 Issues and Trends

The key trend in this sector is integration, moving from different water management functions – treating dirty water, managing floods, gathering and treating water for drinking water, managing land to benefit water management – towards integrating all the needs and functions into IWRM. The legacy in terms of organisations, skill sets, finance, political power and profile and local/national responsibilities can make the transition to IWRM complex, but as the water cycle is itself integrated, it is essential that this change is made globally.

Trends in the sector are helpful markers for future job developments and where you could target your developing knowledge and skills.

Benjamin Tam of Isle Utilities (2020) noted:

> Digitalisation of the water sector continues to be a strong theme and area of rapid progress. From water network monitoring and sewer management, to process optimisation, governance and customer engagement, artificial intelligence (AI) gives the water industry the opportunity to transform itself across all corners and is essential to its quest for better efficiency and greater resilience. . .In terms of water and wastewater treatment, AI and machine learning is being applied to optimise complex and interlinked processes. The aim is to eventually give companies assurance they are striking a precise balance between chemical and power costs and the production of high-quality drinking water or effluent; negating the need to err on the side of caution by over-dosing or over-treating.

3.3   Key Organisations and Employers

As water management (especially IWRM) encompasses many areas of work, there is a correspondingly large number of organisations in the area and every country takes a different approach to their organisational structures.

Global

The Global Water Partnership (founded in 1996 by the World Bank, the United Nations Development Programme (UNDP), and the Swedish International Development Cooperation Agency), promotes integrated water resources management (IWRM) and provides a forum for dialogue among the organisations, government agencies, water users and environmental groups to promote stability through the development, management and sustainable use of water resources.

The network is open to all organisations interested in the management of water resources: government institutions of developed and developing countries, UN agencies, bilateral and multilateral development banks, trade associations, research institutions, non-governmental organisations and the private sector.

The GWP network covers 13 regions: Southern Africa, East Africa, Central Africa, West Africa, the Mediterranean, Central and Eastern Europe, the Caribbean, Central America, South America, Central Asia and Caucasus, South Asia, Southeast Asia and China. The GWP Secretariat is located in Stockholm, Sweden. The network receives financial support from Canada, Denmark, the European Commission, Finland, France, Germany, Netherlands, Norway, Sweden, Spain, Switzerland, UK and USA.

More information can be found on the GWP regional websites.

Engineering and hydrological consultancies and research bodies offer services to assist other organisations in planning and managing water resources. For further information, please see the book chapter on consultancies.

Box 3.1   Key Organisations in the Water Sector

UK

In the UK, the water industry is regulated by the Environment Agency (EA), the Water Services Regulation Authority (Ofwat) and the Drinking Water Inspectorate (DWI). The EA is lead organisation in England for flood risk from main rivers, reservoirs and the sea. Ofwat is the economic regulator of the water and sewerage sectors in England and Wales. Natural Resources Wales have a wide ranging remit in Wales covering flooding, biodiversity, land and water management. SEPA (Scottish Environment Protection Agency) are the key environmental regulator for Scotland and DAERA (Department of Agriculture, Environment and Rural Affairs) have a wide ranging remit in Northern Ireland that includes water, biodiversity and angling.

Water UK is the organisation representing all major statutory water and wastewater service supply organisations in England, Wales, Scotland and Northern Ireland. UKWIR

(Continued)

Box 3.1   (Continued)

---

provides a framework for the procurement of a common research programme for UK water operators on 'one voice' issues.

British Water is an association which has a wide and varied membership covering all sectors of the water industry.

Environmental consultancies are actively involved in water management, delivery and research in the UK.

USA

Federal Emergency Management Agency (FEMA) – coordinates the federal government's role in preparing for, preventing, mitigating the effects of, responding to and recovering from all domestic disasters, including floods.

The Hydrologic Engineering Center (HEC), an organisation within the Institute for Water Resources, is the designated Center of Expertise for the U.S. Army Corps of Engineers in the technical areas of surface and groundwater hydrology, river hydraulics and sediment transport, hydrologic statistics and risk analysis, reservoir system analysis, planning analysis, real-time water control management and a number of other closely associated technical subjects.

Canada

The Institute for Catastrophic Loss Reduction (ICLR) is a world-class centre for multi-disciplinary disaster prevention research and communications, established by Canada's property and casualty insurance industry as an independent, not-for-profit research institute affiliated with the University of Western Ontario.

Natural Resources Canada monitors potential natural hazards and shares information about the risks of natural processes or phenomena that may be a damaging event with Canadians.

Australia

The Australian Emergency Management Institute (AEMI) – is a centre for education, research and training in national emergency management and disaster resilience.

The Australian Institute of Emergency Services (AIES) offers members of emergency service and affiliate organisations the opportunity to be a member of a professional body dedicated to the progression and recognition of the Emergency Service role in the community.

Box 3.2   Professional and Educational Organisations

There are many professional associations and organisations globally which provide valuable information and resources as well as technical education and training for water professionals. They play an essential role in raising and benchmarking professional standards within the sector and are often a helpful networking resource.

UNESCO-IHE Institute for Water Education, the largest international graduate water education facility in the world, is based in Delft, the Netherlands and is part of the UN/UNESCO Water family, being owned by all UNESCO member states. The Institute confers fully accredited MSc degrees, and PhD degrees in collaboration with partner universities.

The Institution of Civil Engineers (ICE) is one of the world's oldest professional institutions, having over 80,000 members in more than 100 countries across the globe. ICE's members are located in many different countries, from Albania to Zambia.

The British Hydrological Society promotes all aspects of hydrology – including the scientific study and practical implications of the movement, distribution and quality of fresh water in the environment.

In Canada, floods are recognised as the most common, largely distributed, natural hazard to life, property, the economy, community/industry water systems and the environment. FloodNet – an NSERC Canadian Strategic Network provide training of highly qualified personnel to develop new knowledge and technology to enhance the capacity for the management and forecasting of floods in Canada. The Network's aims of increasing Canada's knowledge and experience on flood impacts and management through carefully designed projects will not only produce much needed scientific and technical outcomes but also boost the number of highly qualified personnel in hydrology and water resources. In addition to being trained by leading scientists in these fields, highly qualified personnel (HQPs) will be given the opportunity to work on practical water resources problems alongside specialists from the main agencies responsible for hydropower management, policy making, environmental prediction and weather forecasting.

In USA, the Association of State Floodplain Managers (ASFPM) has established a national program for certifying floodplain managers. This program recognises continuing education and professional development that enhances the knowledge and performance of local, state, federal and private-sector floodplain management professionals.

In Australia, Floodplain Management Australia (FMA), formerly the Floodplain Management Association, promote public awareness of flood issues, and provide professional development for floodplain managers. The Association's aim is to encompass promotion of the wise management of floodplains, provision of professional development and information sharing opportunities and representation of the interests of Local Government at State and Federal levels and provide access to a valuable network of practitioners with similar interests and responsibilities in floodplain management.

3.4   Careers in Water Management

Water management professionals are employed in many types of organisations:

- Government agencies & departments
- Environmental and Engineering Consultancies
- Local Authorities

- Infrastructure, network utilities and asset owners
- Catchment managers, Rivers Trusts and environmental NGOs
- Private water, sewage and sanitary utility services companies
- National park and land management authorities
- Science and research bodies
- Academia
- Self-employed individual contractors or principals of start-up firms
- Environmental law firms
- Public health bodies
- Agricultural and Industrial companies (e.g. steel, gas, coal, oil) who use water resources, or whose industry impacts water resources, in order to comply with best management practices, as well as water quality and other environmental regulations
- Specialist campaigning and pressure groups
- The Media

3.5   Job Titles in the Sector

Water Resources – Including Hydrology

Water Resources Director, Water Resources Engineer, Water Resources Analyst, Water Hygiene Consultant, Water Treatment Consultant, Water and Wastewater Treatment Engineer, Water Sales Engineer, Ecologist, Environmental Engineer, District Manager, Building and Grounds Specialist, Risk Analyst, Chemist, Water Manager, Irrigation Engineer, Irrigation Specialist, Pump Station Operator, Geologist, River Modeler, Water Reclamation Specialist and Field Irrigation Manager.

Water Supply and Drinking Water Quality and Wastewater Management

Project Manager, Civil Engineer, Hydraulic Engineer, Mechanical Engineer, Hydrologist, Supervising Engineer, Microbiologist, Reservoir Manager, Construction Manager, Electrician, Pipe Fitter, Surveyor and Laboratory Technician.

Flood and Coastal Erosion Management

Civil Engineer, Hydrologist, Flood Risk Engineer, Flood Modeller, Flood Risk Consultant, Flood and Coastal Risk Management Specialist, Hydraulic Modeller, Flood Risk and Drainage Consultant, project manager, Coastal Modeler, Engineer, Flood Risk Assessor, Flood Risk Manager and Flood Risk Consultant.

IWRM

Other examples of careers within the sector are: Soil Scientist, Wildlife Biologist, Environmental Engineer, Environmental Restoration Planner, Public Affairs Representative, Education Resources Specialist, Government Relations Specialist, Attorney, Regulatory Compliance Specialist, Policy Analyst, Media Strategist and Public Administrator.

3.6   Educational Requirements

The Integrated Water Resources Management sector has such a variety of roles that it is hard to outline firm expectations for educational requirements.

- Entry level jobs may require a high school education and on-the-job training only.
- Technical certification programs may last several months or require two-year associate degrees from community or technical colleges.
- Higher level and knowledge-specific jobs will require bachelor and/or graduate degrees from a college or university which often include Geography, Biological Science, Physical Science, Hydrology, Sustainability, GIS, Environmental Engineering, Geology and Biochemistry.
- Flood risk management is now seen as such an important sector of civil engineering that universities run stand-alone degree courses in the subject.
- A stronger academic background in Hydrology including a BSc and/or MSc in Civil Engineering, Hydrology or Environmental Management may assure a student a better starting position, salary and advancement potential.
- Some programs at the graduate level offer a combination of education and experience to their students by providing them with opportunities to work in the field with practicing non-university water resources professionals.
- A Master of Science (MSc) degree or Doctor of Philosophy (PhD) degree may assure a student a better starting position, salary and advancement potential. In most cases, a PhD is required for a position in academia.
- Many colleges and universities offer short-term certificates and executive education programs for water industry professionals. These programs can also be taken to enhance or update an existing degree or be used as continuing education credits.
- A career in floodplain management may require a Civil Engineering background, whereas management of a major aquifer may require a background in Groundwater Hydrology.
- Many employers will require water resource specialist candidates to have at least two years of related work experience as an engineer, hydrologist, or geologist.
- Working an internship as a student is a great way to gain in-field experience and get your foot in the door with an organisation.
- Even with a completely unrelated first degree it is possible to develop a career in the sector. Experience gained through voluntary work will sometimes compensate for lack of relevant academic background. A vast range of organisations offer voluntary opportunities. Many people working in the sector started with voluntary work, building up contacts, getting known, networking. Paid casual work is rarely advertised and will often be offered to those who have already shown commitment, enthusiasm and hard work as a volunteer. These could include office skills and computer skills.
- In line with increased integration of the sector, water resources planners and managers have found it increasingly necessary to have at least some knowledge of other disciplines including sociology, economics and law. For example, a career in watershed planning, especially in a developed watershed, may require the planner to deal not only with water resources issues but also with biological issues, and with social, economic, or legal issues of the human population of the watershed. The planner must understand the impact of all issues as they relate to the conservation or development of the water resources of the watershed.

3.7 Personal Attributes and Skill Sets

There are many skill sets that will be of use within the IWRM sector. Many skills are developed on the job or through specific training courses. Skills and personal attributes may include:

- Being personable and practical in order to understand the complex relationship between the natural world and human enterprise
- A diverse range of technical and scientific knowledge and skills
- IT skills such as word processing, use of spreadsheets, statistics and graphics packages are essential and knowledge of GPS, GIS, data analysis and integration may be helpful
- Depending on the role, knowledge of physics, soil science, oceanography, geophysics, chemistry, calculus, water quality, hydrology, hydraulics, meteorology and ecology is very good to have in this field
- Good communication skills, teamwork and leadership qualities
- An understanding of relevant legislation, standards and codes of practice
- Development of conceptual and numerical models of the hydrological cycle
- Design and development of plans using industry standard methodologies
- Feasibility and option analysis
- Technical and commercial reporting
- Public consultation and presentation
- Options appraisal including cost estimation
- Programme and/or project management
- Assessment of uncertainty and risks including extreme events, e.g. floods and droughts
- Root cause analysis and problem solving
- Analytical review of peers' work
- Advising on best management of water in its natural state
- Administration of contracts and tender processes on behalf of clients
- Coordination of multi-disciplinary teams
- Conflict resolution
- Liaising with client representatives, local contractors, government agencies, local authorities and suppliers
- Review and provide comments on draft documents, strategies and plans
- Prepare press releases and other media-facing materials
- Computer modelling and design packages
- Risk Management

Training

There are many organisations offering training within the IWRM sector, which include:

American Water Works Association
www.awwa.org/Events-Education/eLearning-Courses

American Society of Civil Engineers
www.asce.org/continuing-education/water-treatment-certificate-program/

International Water Association
www.iwa-network.org/iwa-learn/

International Water Centre
www.watercentre.org/study/courses/

Water New Zealand
www.waternz.org.nz/training

Cranfield University short courses
www.cranfield.ac.uk/courses/short/water/water-short-courses

Water Management Society
www.wmsoc.org.uk/

Develop Training
www.developtraining.co.uk/training/water-and-environmental

Institute of Water UK Water Industry Course
www.instituteofwater.org.uk/uk-water-industry-course/

Water Engineering and Development Centre
www.lboro.ac.uk/research/wedc/

Water Institute of Southern Africa (WISA)
www.wisa.org.za/cpd-training-courses/

Singapore Water Association
www.swa.org.sg/category/swa-training/

Singapore Water Academy
www.pub.gov.sg/sgwa/about

The Water Academy
www.thewateracademy.co.za/training-qualifications

3.8   Career Paths and Case Studies

Personal Profile

Dr Alexandros Stefanakis, Assistant Professor, Technical University of Crete and Constructed Wetland Consultant, Crete

I am an Environmental Engineer with a Master in Hydraulic Engineering and a Ph.D. in Ecological Engineering. I am specialised in water engineering, focusing on wastewater treatment using natural systems such as Constructed Wetlands.

Currently I am Assistant Professor at the School of Environmental Engineering, Technical University of Crete in Greece and Visiting Lecturer at the German University of Technology in Oman. In the past I worked not only in various research Institutes and

Universities in Europe (Germany, UK, Portugal), but also in the private sector as Specialist in the design of municipal and industrial wastewater treatment plants using nature-based solutions.

For many people, working in academia is quite different from working in the private sector or in the industry. Although this is not totally wrong, both sectors have some basic common requirements and skills. First of all, the ability to work in teams and collaborate with colleagues from different disciplines and often from different countries is necessary. Teamwork is always what makes the best outcome no matter the nature of the task/project. Organisational and management skills are also very important in order to be able to manage a continuous flow of data and information and also to coordinate the work carried out between different teams. To this, good communications skills are also needed; especially in the private sector when you frequently come into contact with clients, but also in the academia when you are delivering lectures or presentations, or you are requesting funding for research.

Of course, in all cases, excellent background and knowledge of the relative scientific field is the basis for professional development. Continuous, lifelong education is necessary in order to keep up with the latest developments and state-of-the-art technological improvements in water sector. And this can be achieved only through personal efforts. Quite often I see younger professionals or students who believe that after their graduation or once they get their first job position, they are at the end of their path. In fact, this is only the beginning. I can tell that as a graduated and professional I am studying more than I used to as a student.

I had also the chance to work in two, supposedly, different sectors; the academia and the industry. Academics often possess a very good knowledge of the theoretical background but sometimes lack in practical experience. I believe that having the appropriate scientific background and also by gaining experience in the actual implementation of, i.e., a water technology is what makes an individual successful and represent the best credential for a continuous personal and professional development. After all, coming into contact with the industry and private clients you get to understand better the actual problems they face, which respectively could help you to focus your research goals on these issues. And this is exactly the link between academia/research and the industry, since the academic sector can focus on research fields, topics and applications that the industry has identified and for which improved solutions and knowledge are needed.

Overall, water demand is rising while available water resources are limited. Thus, new technologies or upgrading existing technologies are needed to provide clean water and prevent pollution.

Therefore, I believe that the water sector and especially sustainable technologies for water treatment will continue to meet noticeable development in the following years. A career in the water sector is a demanding one. It requires continuous learning and advanced management and communication skills. But it can also be profitable and satisfying.

Source: Dr Alexandros Stefanakis, Assistant Professor, Technical University of Crete and Constructed Wetland Consultant, Crete. © John Wiley & Sons.

Personal Profile

Michael Rhodes, Mott MacDonald, United Kingdom

Following a very wide-ranging degree in Environmental Science and an enjoyable year spent working in a small environmental team in the public sector, I wanted to find out what working in the environmental consultancy sector was all about. Mott MacDonald have a highly respected name in consultancy with exciting projects taking place across the globe, the chance to work with and learn from experts in their field sounded like a great first step on the career ladder. I liked the idea of varied project work with an important interdisciplinary and collaborative focus, so I knew a career with Mott MacDonald would be engaging and enjoyable place to work.

My first exposure to Mott MacDonald was through careers fairs and talks organised by my university and other professional groups (these are surely the best way for students to expand their horizons and find out what they could be doing in a few years' time!). I met a fantastically friendly and enthusiastic hydro-geologist at one of these events who shared her career progress and dutifully listened to mine at the careers fair each year, four years later I was still talking to the same familiar face and I knew Mott MacDonald would be a great place to work.

Source: Michael Rhodes, Mott MacDonald, United Kingdom. © John Wiley & Sons.

Personal Profile

Abdulhamid Gwaram, Project Officer, Government Small Towns Water Supply & Sanitation program, Nigeria

After studying Civil Engineering in Nigeria, I worked with the water supply department of the federal ministry of water resources for over a decade as a civil/water resources engineer.

I was involved in many projects that apply principles of Integrated Water Resources Management (IWRM), such as the involvement and empowerment of stakeholders, especially women and local users, in decision-making in the water sector.

I was therefore very excited when I saw the opportunity to study the International Water Centre's (IWC) Master of Integrated Water Management in Australia in 2011 under the AusAID scholarship, and I made my choice without hesitation.

This scholarship helped me realise my dream of obtaining international level knowledge of water management. It was also an opportunity for me to learn how things are done in other places, and to make great new friends from all over the world.

I am now the project officer in charge of implementation of the Small Towns Water Supply and Sanitation program, improving the water supply coverage of small towns in Nigeria in a sustainable manner.

Source: Abdulhamid Gwaram, Project Officer, Government Small Towns Water Supply & Sanitation program, Nigeria. © John Wiley & Sons.

Personal Profile

Russell Martin, Independent Researcher, USA

I am the Drought Monitor Focal Point for the 12 State Northeast Region of the USA. Since drought is a multifaceted phenomenon which exists on multiple temporal and spatial scales, and whose impacts are not only determined by precipitation amounts but also the interaction of precipitation with society and the hydrologic system, I monitor data from many different sources in order to assess the existing level of drought in my geographical region. The data include, but are not limited to, precipitation measurements from rain gauges, radars and satellites, stream levels, soil moisture measurements, crop and other vegetative conditions, water well levels, snow pack amounts and media reports of drought impacts such as water supply shortages.

I also collaborate with local experts, including State Climatologists and meteorologists and hydrologists at local National Weather Service offices, to incorporate their appraisals of conditions in their areas into my recommendations for the regional drought depiction on the weekly United States Drought Monitor.

To do this job I use skills from my background in meteorology and climatology, especially statistics and data quality assurance. My knowledge of hydrology, water resources and agriculture are also useful, along with the ability to communicate clearly yet diplomatically when disagreements sometimes arise about drought conditions for which the data show conflicting indications or admit alternate interpretations.

I came to this job by a long and winding road from theoretical physics through satellite meteorology to climate forecasting and applied climatology, with a couple of software engineering jobs sprinkled in.

At the present time in the U.S., many people working on drought are doing so as only part of their jobs, so people desiring to work in the field should acquire multiple skills which qualify them to do a variety of climate, weather, or water resource related tasks. In particular, GIS skills are becoming increasingly important; in fact, almost a necessity.

Source: Russell Martin, Independent Researcher, USA. © John Wiley & Sons.

Personal Profile

Andrew M., Flood and Coastal Risk Management Advisor, Environment Agency, UK

On completing his A-levels, Andrew knew that he didn't want to go to university. He thought about joining the police and spent some time as a volunteer constable, but his love of geography and the outdoors led him to the Environment Agency. He joined the Environment Agency on their Foundation Degree training scheme and is now gradually developing both his career and academic qualifications. He feels he has found a job that he loves doing and enjoys the variety of time in the office and out wading in rivers!

Source: Andrew M., Flood and Coastal Risk Management Advisor, Environment Agency, UK. © John Wiley & Sons.

3.9   External Resources

UK

Organisation	Website
Environment Agency (EA)	www.gov.uk/government/organisations/environment-agency
Natural Resources Wales	www.naturalresources.wales
Scottish Environment Protection Agency	www.sepa.org.uk
Department of Agriculture and Rural Development	www.dardni.gov.uk
Chartered Institution of Water and Environmental Management (CIWEM)	www.ciwem.org
Water Management Society (WMSoc)	www.wmsoc.org.uk
Centre for Ecology & Hydrology (CEH)	www.ceh.ac.uk
Cranfield Water Science Institute (CWSI)	www.cranfield.ac.uk
British Hydrological Society	www.hydrology.org.uk
Institution of Civil Engineers (ICE)	www.ice.org.uk
	www.ice.org.uk/careers-and-training/return-to-a-career-in-civil-engineering/which-civil-engineering-role-is-right-for-you/career-profile-water-resources-engineering
Centre for Water and Development SOAS	www.soas.ac.uk/centres
Ofwat	www.ofwat.com
Water UK	www.water.uk.org
British Water	www.britishwater.co.uk
Foundation for Water Research	www.fwr.org/
UKWIR	www.ukwir.org
Environmental Change Institute (ECI) and Oxford University Environmental Change and Management flagship MSc/MPhil programme aims to train international leaders	www.eci.ox.ac.uk/teaching/index.php
Flood Risk Management Research Consortium (UK)	www.floodrisk.org.uk

Europe

Organisation	Website
European Water Association (EWA)	www.ewa-online.eu
European Union of Water Management Associations (EUWMA)	www.euwma.org
European Water Resources Association (EWRA)	www.ewra.net
Water Reuse Europe (WRE)	www.water-reuse.eu

Organisation	Website
Global Water Partnership – Central and Eastern Europe	www.gwp.org/en/gwp-in-action/Central-and-Eastern-Europe
Global Water Partnership – Mediterranean	www.gwp.org/en/gwp-in-action/Mediterranean

Asia

Organisation	Website
International Water Resources Association (IWRA) Japan Committee	www.iwra.org
International Water Resources Association (IWRA) China Committee	www.iwra.org
International Water Resources Association (IWRA) India Committee	www.iwra.org
International Water Management Institute (IWMI) – Central Asia	www.centralasia.iwmi.cgiar.org
International Water Management Institute (IWMI) – South Asia	www.southasia.iwmi.cgiar.org
International Water Management Institute (IWMI) – Southeast Asia	www.sea.iwmi.cgiar.org
Global Water Partnership – Central Asia and Caucasus region (CACENA)	www.gwp.org/en/gwp-in-action/Central-Asia-and-Caucasus
Global Water Partnership – China	www.gwp.org/en/gwp-in-action/China
Global Water Partnership – South Asia	www.gwp.org/en/gwp-in-action/South-Asia
Global Water Partnership – Southeast Asia	www.gwp.org/en/gwp-in-action/Southeast-Asia

Africa

Organisation	Website
International Water Management Institute (IWMI) –East Africa	www.eastafrica.iwmi.cgiar.org
International Water Management Institute (IWMI) –West Africa	www.westafrica.iwmi.cgiar.org
International Water Management Institute (IWMI) – Southern Africa	www.southernafrica.iwmi.cgiar.org
Global Water Partnership – Central Africa	www.gwp.org/en/gwp-in-action/Central-Africa

(Continued)

Organisation	Website
Global Water Partnership – Eastern Africa	www.gwp.org/en/gwp-in-action/Eastern-Africa
Global Water Partnership – Southern Africa	www.gwp.org/en/gwp-in-action/Southern-Africa
Global Water Partnership – West Africa	www.gwp.org/en/gwp-in-action/West-Africa

North America – Canada

Organisation	Website
Canadian Water Resources Association (CWRA)	www.cwra.org
Canadian Water and Wastewater Association (CWWA)	www.cwwa.ca
Water Environment Federation (WEF)	www.wef.org www.wef.org/jobbank
Alberta Water & Wastewater Operators Association (AWWOA)	www.awwoa.ab.ca
Manitoba Water & Wastewater Association (MWWA)	www.mwwa.net
Northern Territories Water and Wastewater Association (NTWWA)	www.ntwwa.com
Saskatchewan Water and Wastewater Association (SWWA)	www.swwa.ca
Western Canada Water Association (WCWA)	www.wcwwa.ca
FloodNet NSERC	www.nsercfloodnet.ca
Toronto and Region Conservation Authority	www.trca.on.ca
Institute for Catastrophic Loss Reduction (ICLR)	www.iclr.org
Natural Resources Canada monitors potential natural hazards and shares information about the risks of natural processes or phenomena that may be a damaging event with Canadians	www.nrcan.gc.ca

North America - USA

Organisation	Website
Federal Emergency Management Agency (FEMA)	www.fema.gov
U.S. Army Corps of Engineers (USACE)	www.usace.army.mil
Institute for Water Resources (IWR)	www.iwr.usace.army.mil
The Hydrologic Engineering Center (HEC),	www.hec.usace.army.mil
American Institute of Hydrology (AIH)	www.aihydrology.org
American Water Resources Association (AWRA)	www.awra.org

Organisation	Website
American Water Works Association (AWWA)	www.awwa.org
International Desalination Association (IDA)	www.idadesal.org
National Water Resources Association (NWRA)	www.nwra.org
National Ground Water Association	www.ngwa.org
National Rural Water Association (NRWA)	www.nrwa.org
Water Environment Federation (WEF)	www.wef.org
Water Quality Association (WQA)	www.wqa.org
Association of State Floodplain Managers (ASFM)	www.floods.org
Water District Jobs	www.waterdistrictjobs.com

South America

Organisation	Website
Water Centre for the Humid Tropics of Latin America and The Caribbean (CATHALAC)	www.cathalac.int/en
Water Center for Arid and Semi-Arid Zones in Latin America and the Caribbean (CAZALAC)	www.cazalac.org
Global Water Partnership – Caribbean	www.gwp.org/en/gwp-in-action/Caribbean
Global Water Partnership – Central America	www.gwp.org/en/gwp-in-action/Central-America
Global Water Partnership – South America	www.gwp.org/en/gwp-in-action/South-America

Oceania

Organisation	Website
Australian Water Association (AWA)	www.awa.asn.au
Floodplain Management Australia (FMA)	www.floods.org.au
International Water Resources Association (IWRA) Oceania Committee	www.iwra.org
Australian Disaster Resilience Knowledge Hub	www.knowledge.aidr.org.au/about
Australian Institute of Emergency Services (AIES)	www.aies.net.au
Hydrological Society of South Australia Inc.	www.hydsoc.org
Water Services Association of Australia (WSAA)	www.wsaa.asn.au
Water Industry Operators Association of Australia (WIOA)	www.wioa.org.au
Water New Zealand	www.waternz.org.nz/About
National Institute of Water and Atmospheric Research (NIWA)	www.niwa.co.nz

Global

Organisation	Website
International Water Resources Association (IWRA)	www.iwra.org
The International Water Management Institute (IWMI)	www.iwmi.cgiar.org
The International Water Association (IWA)	www.iwa-network.org
UNESCO-IHE Institute for Water Education	www.unesco-ihe.org
Josh's Water Jobs	www.joshswaterjobs.com/jobs

FOR UPDATED AND ADDITIONAL RESOURCES, INCLUDING EXTRA CHAPTERS, GO TO WWW.ENV.CAREERS

References

Bouwer, H. (2000). Integrated water management: emerging issues and challenges. *Journal of Agricultural Water Management* 45 (1): 217–228.

Ecosystems Knowledge Network (2015). Managing river catchments. www.ecosystems knowledge.net/resources/tools-guidelines/catchment-management (accessed 7 July 2020).

Global Water Partnership (n.d.). What is IWRM? www.gwp.org/en/GWP-CEE/about/why/what-is-iwrm/ (accessed 5 April 2021).

Jha, A.K., Bloch, R., and Lamond, J. (2012). Cities and Flooding: A Guide to Integrated Urban Flood Risk Management for the 21st Century. World Bank. www.openknowledge.worldbank.org/handle/10986/2241 License: CC BY 3.0 IGO.

Mehta, L. and Movik, S. (2014). Flows and Practices: Integrated Water Resources Management (IWRM) in African Contexts. IDS Working Paper 438 First published by the Institute of Development Studies in February 2014.

Tam, B., Isle Utilities (2020). Harnessing the full benefit of AI in water. LinkedIn article Published on 26 March 2020.

UNESCO (n.d.). World Water Development Report. www.unesco.org/new/en/natural-sciences/environment/water/wwap/wwdr/ (accessed 7 July 2020).

UN Water (n.d.). Water quality and wastewater. www.unwater.org/water-facts/quality-and-wastewater/ (accessed 7 July 2020).

World Health Organisation (WHO) (n.d.a). 'Drinking water fact sheet. www.who.int/news-room/fact-sheets/detail/drinking-water (accessed 7 July 2020).

WHO (n.d.b). Health topics: floods. www.who.int/health-topics/floods (accessed 7 July 2020).

4

Environmental Law

4.1   Sector Outline

Environmental law is a vast practice area which encompasses a broad range of substantive areas of law including administrative law, tort law (related to harms caused to others), criminal and regulatory law, constitutional law and property law and has expanded to include international environmental governance, international trade, environmental justice and human rights and climate change.

Only a few decades ago environmental law was not considered a vital area of international and national concern and was not a high priority on the global agenda. In recent years however, growing recognition that we are on the verge of an ecological and climatic crisis with irreparable consequences has pushed environmental law and policy to the forefront of public and media concern. This has forced governments and corporations to pivot and start to more coherently address environmental law and policy issues. Those working in the environmental law field tackle a wide range issues and problems such as sustainable resource development, contamination of land and water, health and safety, recycling and waste management, international trade, disaster management and assistance with development of alternative or green energy sources (All About Law, 2015).

Most modern environmental laws started to be enacted in the late 1960s and early 1970s. For example, in the 1970s member countries of the EU introduced laws to ensure the cautious use of natural resources, to minimise the ecological effects of consumption and production with particular reference to waste, conserve biodiversity and areas of key biological importance (EUR-Lex, 2015).

Similarly, in the USA, numerous foundational federal and state statutes were passed into law and implemented in the 1970s. These included the Clean Air Act, the Clean Water Act, the Safe Drinking Water Act, the Comprehensive Environmental Response Compensation and Liability Act (CERCLA), the Toxic Substances Control Act, the National Environmental Policy and the Endangered Species Act of 1973, which designated and protected species on the verge of extinction. However, these laws were views as taking a 'silo' approach and failing to recognise the interconnections between water quality, watershed and wetland development and protection of wildlife habitat.

Global Environmental Careers: The Worldwide Green Jobs Resource, First Edition. Justin Taberham.
© 2022 John Wiley & Sons Ltd. Published 2022 by John Wiley & Sons Ltd.

The landmark report that began to shift awareness on the need to develop more sound and comprehensive environmental law and policy was the Brundtland Commission's pathbreaking report, Our Common Future (Brundtland 1987). One of the primary findings of the Brundtland Commission report is that promoting sustainability requires environmental, social and economic considerations to be fully integrated into government, corporate, community and institutional policy and decision-making processes. (McRobert and Ruby 2008) Many of the approaches to improving these decision-making processes that have been undertaken in the past 30 years have proven inadequate. Consequently citizens, corporations and other actors (such as civil society organisations) are challenging international, national and sub-national governments and administrative systems to provide creative solutions to our urgent problems in environmental and resource management and to begin to develop a new synthesis that responds to international treaties such as the Paris Agreement on Climate Change (United Nations 2015) and other similar regional, bioregional and national initiatives.

Many advocates working for civil society organisations argue that environmental law should be primarily concerned with the ever-increasing 'footprint' humans are leaving on the planet. Further, they contend that this has been caused by processes such as industrialisation and urbanisation, overexploitation of natural resources (the oceans and rainforests included) and unprecedented levels of pollution (The University of Law 2013). Other lawyers working for governments and corporations argue that environmental laws, regulations and policies are intended to ensure that health and safety is safeguarded, resources are managed sustainably and parks and ensure wilderness areas are adequately protected.

Particular challenges in implementing sustainability include: (i) declining budgets and reduced investigation and enforcement for environmental matters; (ii) deficiencies in information available to guide decision-makers; (iii) poor transparency and accountability in decision-making about environmental values and science; (iv) weak engagement of the public in participation; and (v) inability to access courts and tribunals to address these issues listed above (McRobert and Ruby 2008).

In the wake of the Brundtland report, most environmental lawyers began to recognise the enormous potential scope of environmental law and policy. Consequently, lawyers in the sector have become more specialised and often tend to focus on particular areas of expertise such as water law, environmental impact assessment, international treaties, air approvals, etc. and target provision of their services to companies and government agencies.

In the past 15–20 years, environmental law has been developing at a rapid pace in South America, particularly in regions such as Brazil, Argentina and Peru and is currently more sophisticated and well managed than ever before, often mirroring those protocols initiated in North America and the EU (Beveridge, and Diamond, P.C 2016).

In Asia, the demand for those working in environmental law is growing in countries such as Indonesia who are currently experiencing an emerging conflict between regional laws and sectoral laws who focus on discouraging natural resource depletion such as deforestation through illegal logging (Tan, 2004). Asian countries with more comprehensive environmental legislation quickly establish an environmental lawyer capacity to match.

As far back as 1999, Prof Ben Boer noted 'More recently, in many non-Western countries and especially in Asia, environmental law has begun to enter into adulthood, manifested by significant legislative initiatives, judicial activism and a resulting environmental jurisprudence, and the establishment and growth of environmental and resource management agencies.'

Harashima (2000) assessed governance in Asia's developing economies:

> In Asian countries, many positive trends can be found in environmental governance. Environmental laws were strengthened, particularly in the 1970s and again in the 1990s. [However] environmental governance systems in Asia have not yet developed satisfactorily at the national level.

Zhao, in his Zhao 2019 working paper for the Asian Development Bank Institute highlighted Asian challenges:

> Asia faces daunting environmental challenges. A recent UNEP report finds that about 4 billion people or over 92% of the population in Asia and the Pacific are exposed to air pollution exceeding or far exceeding WHO guidelines (United Nations Environment Program (UNEP) 2019). Some of the challenges arise from rapid industrialisation, as evidenced by the severe air, water and soil pollution in the People's Republic of China (PRC) and India, and some arise from economic development leading to resource degradation, e.g., deforestation and loss of biodiversity in Southeast Asia. . . Many developing countries lack adequate legal systems for citizens to bring civil or criminal lawsuits against polluters. Furthermore, corruption often undermines the incentives of government officials to enforce environmental laws and regulations.

In support, Raine and Pluchon (2019) noted:

> Effective laws, coupled with empowered institutions and citizens to ensure their implementation, provide the critical enabling environment necessary to deliver the environmental dimensions of the 2030 Agenda for Sustainable Development and other internationally agreed environmental goals. While most countries have now developed environmental legal frameworks at different levels, effective implementation remains a key challenge for almost all countries. In addition, many environmental legal frameworks need to be strengthened to respond to new and emerging issues, including new internationally agreed goals and commitments. The United Nations Environment Programme (UN Environment), working with partners, supports countries around the world to meet these objectives and needs. A central focus of its work is to support countries to advance the environmental rule of law.

In terms of Africa, Alexander Paterson, in Kameri-Mbote et al. (2019) highlighted:

> In keeping with the spirit of forging African solutions to African challenges, scholars from across the continent formed the Association of Environmental Law Lecturers from African Universities (ASSELLAU) in 2004. Its specific objectives include promoting the generation and dissemination of environmental law research to assist Africa's law and policymakers to craft and implement legal frameworks that

achieve the tricky balance referred to above through conferences and symposia. In partial fulfilment of this mandate, ASSELLAU held its 4th Scientific Conference in Yaoundé, Cameroon, from 10–13 January 2018.

The Global Environmental Outlook GEO-6 – Regional Assessment for Africa (UNEP 2016) highlighted:

> Africa faces both enormous challenges in relation to environmental management, and equally huge opportunities for 'doing this better'. Africa's natural capital is challenged by competing uses, illegal off-take, weak resource management practices, climate change and pollution. This calls for forward looking, flexible, inclusive and integrated approaches in the formulation and implementation of policies.

Practice Settings

The three most common settings in which public interest environmental lawyers practice are government, non-profit organisations and private public interest firms.

- National/Federal Government – key organisations include Government departments and Government bodies which may control land management and environmental protection
- State and Regional Governments – key organisations include more local equivalents of the national bodies and the relevant bodies involved in climate change policy. In addition, local governments may employ environmental lawyers and some large urban areas have municipal environmental agencies
- International bodies – environmental lawyers are employed by bodies such as the United Nations and the World Bank
- Non-Profits – lawyers for these organisations are often involved in litigation, lobbying, education and negotiation. Examples in the USA include the Environmental Defence Fund and the Natural Resources Defence Council
- Private Public Interest Law Firms – there is a growing number of firms known for their environmental law work

4.2 Issues and Trends

There are a number of key themes in the Environmental Law sector, as outlined by Harvard Law School's 'A Trail Guide to Careers in Environmental Law' (2013).

Climate Change Law

Lawyers in this field must adapt existing environmental regulatory. They amend laws, rewrite regulations, or interpret statutes in new ways. Climate change lawyers also work to address the cause of climate change: greenhouse gas emissions. Designing legal tools to incentivise emissions reduction, like cap-and-trade programs or carbon taxes, is a current priority. Many climate change lawyers, support the development, regulation, permitting and financing of new technologies to reduce emissions.

Energy Law

Government lawyers in this area ensure that companies follow mandated price structures, permit rules and emission standards. Energy lawyers are laying the legal groundwork for cleaner sources of energy. They design regulatory regimes for new power sources or advocate for climate conscious energy policies. They also support the permitting and financing of solar, wind, geothermal and hydroelectric power.

Food Law

The relatively new field of 'food law' aims to remedy problems in our food systems, including lack of access to fresh food, risks from genetically engineered foods and the effects of climate change on agriculture. With the goal of making food systems safer and healthier, government and non-profit lawyers research food problems and inform policymaking. They might draft guidelines, laws and regulations, or educate and empower communities about food issues.

Water Law

Laws protect water sources from pollution and defend marine habitats from degradation. As a result of climate change and environmental accidents, issues in water law are abundant and constantly changing; oil spills, melting polar ice caps and shipping emissions are only a few current problems. A key objective for many water lawyers is to ensure access to safe, local and reliable sources of drinking water. Other water lawyers preserve ecosystems and protect marine species from destructive fishing and climate change.

Human Rights Law

The United Nations (1948) notes:

> The Universal Declaration of Human Rights is generally agreed to be the foundation of international human rights law. Adopted in 1948, the UDHR has inspired a rich body of legally binding international human rights treaties. It continues to be an inspiration to us all whether in addressing injustices, in times of conflicts, in societies suffering repression and in our efforts towards achieving universal enjoyment of human rights.

Ben Boer (2015) highlighted:

> Despite the fact that most countries in the Asian region are members of the global human rights treaties, the actual implementation of legislative and policy frameworks concerning human rights is generally underdeveloped. . .However, despite the inadequacy of the environmental legal frameworks and the lack of government

implementation in some countries, particularly in South Asia and Southeast Asia, the courts have nevertheless been able to achieve significant environmental outcomes by using national constitutional provisions focused on basic human rights, especially the right to life.

4.3 Key Organisations and Employers

The Global 500 website https://www.legal500.com is a valuable independent and impartial resource which outlines global law firm capabilities. It is an important tool you can use to search for jobs at large firms who are retained by larger employers and government agencies working on regulatory issues and large development projects. Government and civil society and environmental group organisation web sites are another way to search for likely employers within the sector.

In the UK, those working in environmental law tend to be first and foremost concerned with property or project finance departments (All About Law, 2015). However, there are a number of firms becoming known for their environmental work. The Legal 500, which analyses law firm capabilities, ranks top firms (London based, Environment, 2020) as:

Box 4.1   Top Law Firms, London based, Environment Sector

Allen & Overy LLP – clients include Horizon Nuclear, HP and Allnex

Burges Salmon LLP – clients include John Lewis Partnership, Affinity Water and BT

Clifford Chance LLP – clients include Coca-Cola, Global Infrastructure Partners and Nielsen Holdings

CMS – clients include Deliveroo, Brett Aggregates and SUEZ

Dentons – clients include PWC, Royal Mail Group and Sainsbury's

Leigh Day – clients include Friends of the Earth, Cerrejon coal mine victims group action, Colombia and Wild Justice

Linklaters LLP – clients include Unilever, Linde and Cummins

The Legal 500, which analyses law firm capabilities, ranks top Barristers and Chambers (London based, Environment, 2020) as:

Box 4.2   Top Barristers and Chambers, London based, Environment Sector

Leading Sets

39 Essex Chambers

Francis Taylor Building

Landmark Chambers

Leading 'Silks'

David Elvin QC – Landmark Chambers

Charles Gibson QC – Henderson Chambers

(Continued)

Box 4.2   (Continued)

David Hart QC – 1 Crown Office Row
Stephen Hockman QC – SIX PUMP COURT
Robert McCracken QC – Francis Taylor Building
Stephen Tromans QC – 39 Essex Chambers

North America has a range of institutions offering employment in the sector of environmental law. For instance, the Environmental Protection Agency provides administration of environmental regulation and work to design laws based on benefiting human life and the natural environment (Beveridge & Diamond, P.C., 2016).

The Legal 500, which analyses law firm capabilities, ranks top firms (US, Environment, 2020) as:

Box 4.3   Top Law Firms, US based, Environment Sector

Litigation
Arnold & Porter – clients include Dow Chemical, Honeywell International and BP
Gibson, Dunn & Crutcher LLP – clients include Starbucks Coffee, Dole Food and Boeing
Latham & Watkins LLP – clients include Monsanto, Dow Chemical Chevron Corporation and California American Water Company

Regulatory
Beveridge & Diamond, P.C. – clients include 3M, Bayer and Chevron
Hunton Andrews Kurth LLP – clients include Cargill, Air Permitting Forum and Marathon Petroleum Corporation

Transactional
Cravath, Swaine & Moore LLP – clients include Avon, Disney and Johnson & Johnson
Kirkland & Ellis LLP – clients include Bristol Myers Squibb Company, Marriott Vacations Worldwide Corporation and American Securities
Latham & Watkins LLP – clients include The Carlyle Group, Nouryon and Platinum Equity
Weil, Gotshal & Manges LLP – clients include Eli Lilly and Company/Elanco, General Electric and Genstar Capital

---

The Australian government has a range of organisations who work to form legislation regarding the environment. One such example is the Environment Protection and Biodiversity Conservation Act 1999 (Australian Government 1999), which acts to provide legal framework regarding the protection and management of nationally and internationally important flora, fauna and ecological communities. Great focus for environmental law workers in Australia surrounds the protection of the Great Barrier Reef and its associated marine protection areas (Australian Government Department of the Environment, 2015).

One potential area for work in the sector of Environmental Law in Africa is with the non-profit Centre for Environmental Rights (2021). This centre was formed in 2009 by eight civil society organisations (CSO's) and aims to work with local communities in South Africa to promote environmental sustainability (Centre for Environmental Rights, 2016).

The Environmental Law Institute has also worked in Africa since 1999 and has been operating in partnership with African NGO's and governments to aid in advancing and forming environmental policies and laws (Centre for Environmental Rights, 2016). ELI's work has included operating in Nigeria to help manage the growing problem of mercury from the processing of gold ore; this can leak into waterways and damage both aquatic plant life and organisms (Goldman, 2013).

4.4   Careers in the Sector

In the Siskinds Blog article 'So you want to be an environmental lawyer. . .', Diane Saxe (2013) offers some helpful and honest advice:

> it is exciting and reassuring that so many talented young people want to devote their professional lives to environmental protection. On the other hand, there are far fewer jobs in our field than there are people who want those jobs. . .The more exciting, 'save the world' cases that students imagine must usually be done pro bono and most of them don't succeed. . .Environmental groups and civil servants more often have the chance to do the cases you probably want to do. But environmental groups usually struggle with limited resources, and often feel they are beating their heads against a very hard wall. . .the best place to learn is often in the public service, where the problems are the most important, and there are many wonderful lawyers and scientists to learn from. No wonder civil service jobs are so hard to get. . .
>
> Even if you end up practicing in another area of law, or working in some other capacity, you will have opportunities to work for environmental protection. Lawyers in any practice area can bring strong analytical skills and an articulate voice, as well as financial support, to the environmental issues that affect their organisation and their community. And sometimes, the job of your dreams will show up years later, if you keep the door open while you're building your skills.

An ideal way to get to know the sector is to join the student section of the relevant professional or trade body – often there are organisations solely for environmental lawyers, such as the UKELA in the UK. These bodies organise talks and networking events and are an ideal way to get to meet people working in the sector and this could lead to internships or work experience.

Volunteering in the sector is valuable – such as getting involved in writing papers and campaign materials at 'Legal Aid' bodies and environmental charities.

4.5   Job Titles in the Sector

Attorney, Lawyer, Barrister, QC, Partner, Paralegal, Researcher, Administrator.

4.6   Educational Requirements

ClientEarth (2019) outlined the key routes to becoming an environmental lawyer in their article 'How to become an environmental lawyer in the UK':

> In the UK, there are typically two routes you can follow to study to become an environmental lawyer. One option is to read a full law degree (LLB) as an undergraduate degree. In doing so you learn about the 'core' components of the law and legal system as well as electives of a range of legal topics, one or many of which could be environmental law or similar disciplines.
>
> Alternatively, you could pursue an undergraduate degree of a different discipline and, after graduating, undertake a one-year 'law conversion'. This is formally known as a Graduate Diploma in Law (GDL) or a Common Professional Examination (CPE) which focuses on teaching the core components of the law and legal system in a year.
>
> After completing your studies, you will need to complete professional education. In England and Wales the legal profession is divided into two: solicitors and barristers. There are distinct courses depending on which path you wish to follow. The barrister profession is regulated by the Bar Standards Board. Solicitors are regulated by the Solicitors Regulation Authority.

The website environmentalscience.org (2021) outlines the educational requirements in the USA:

> To practice as an Environmental Lawyer, you will need to obtain a law degree commonly known as a Bachelor of Laws (LLB) or a Juris Doctor (JD). The JD is a postgraduate qualification, the bachelor is an undergraduate qualification. Further qualification is also available including a Masters of Law (LLM), Masters of Environmental Law, or a Doctor of Philosophy (PhD). These further options are a good option for lawyers seeking more senior roles or a higher pay grade.
>
> Once you have completed your law degree, you are required to gain practical experience and sit further examinations to be admitted to the bar. Once you are admitted to the bar you can represent clients and formally provide legal advice.

4.7   Personal Attributes and Skill Sets

The following skills are needed:

- Effective oral and written communications
- Clearly getting points across and to create a strong argument
- Good interpersonal skills
- Digesting detailed information and having an eye for detail
- Reciting case details
- Handling delicate details and issues
- Teamwork

- Honesty and integrity
- Ambition
- Knowledge in the sciences is helpful in dealing with technical cases and discussing details with scientists and researchers

4.8 Career Paths and Case Studies

Personal Profile

David McRobert, Environmental and Indigenous Rights Lawyer and Author, Peterborough, Ontario, Canada

I became passionate about environmental and energy issues in the early 1970s at the age of 13. The control OPEC was exerting on energy supplies to developed nations led to the acceleration of the development of nuclear power, hydroelectric dams, new coal and natural gas plants.

I wanted to help bring about the social transformations that are necessary to help Canada become a more sustainable society. I have been guided in my law reform advocacy, policy work and activism by the idea that, 'if we don't shape our world, someone else will and we probably won't like the results.'

In 1973 the rubber hit the road for me big time when the Quebec government proposed a sweeping plan to develop hydroelectric dams across many of the majestic rivers in the James Bay region of northern Quebec. I became mesmerised by the intense public policy debate that emerged as the Cree Indians and Inupiat residents challenged the right of the Quebec government to launch the projects. The James Bay Hydro project dispute highlighted to me that Canada's Aboriginal peoples had a vital and compelling voice on large energy and mining projects, one that they continue to exercise with legitimacy and passion.

Subsequent projects – such as the Mackenzie Valley Pipeline proposal (1970) in the Northwest Territories and massive 'oil sands' extraction expansion in the 1990s and 2000s in Alberta and related recent pipeline construction and port delivery projects in British Columbia – reinforced my perception on this.

In 1977 I undertook a B.Sc. in Biology at Trent University in Peterborough, Ontario and I was accepted to study medicine at several Canadian universities. After a series of personal events and studying in Europe, in 1981 I undertook a Master's degree in Environmental Studies at York University, focusing on biological conservation in the Canadian Arctic.

I worked in the Western Arctic, Mackenzie Delta and Yukon to gain practical experience and conduct field research for my master's thesis. By the time I completed my MES degree, I knew that the Inuit and Yukon Indians didn't need more doctors. They needed lawyers with science backgrounds to help them to safeguard their rights and access to resources and protect northern landscapes from rapid oil and gas and mining development projects.

During my law education and training between 1984 and 1989, I began working part-time, advocating, volunteering and writing magazine and newspaper articles with the numerous environmental groups and the Green Party of Canada and Pollution Probe. I also worked at and wrote magazine articles for Pollution Probe in Toronto (then Canada's largest environmental organisation). This paid off – in 1989 I became a full-time campaigner and advocate on waste reduction and climate change at Pollution Probe. I attended United Nations climate change conferences and worked on waste reduction policies and air pollution.

Probably my biggest career impact has been in the field of waste reduction and recycling. Between 1990 and 1993 I worked at the Ontario Ministry of the Environment on drafting the laws and regulations that required most municipalities to provide curbside recycling programs and yard waste collection to residents and promoting product stewardship for packaging and the circular economy.

In late 1994 I joined the office of the Environmental Commissioner of Ontario (ECO) where I worked as the senior lawyer for 16 years, auditing the environmental compliance performance of 16 government ministries and agencies. Consequently my 'fingerprints' are on a wide range of law and policy reforms in Ontario in the late 1990s and through the 2000s, including the Safe Drinking Water Act, and laws around aggregates, endangered species, wildlife management and wetlands and wilderness protection.

In 2010, I left the ECO and began work in the private sector. In this capacity I have worked representing a wide range of indigenous, individuals and private sector clients on various issues through written and oral advocacy and in courts, at tribunals, mediation hearings and in other forums.

My work experience and academic background on environmental policy has inspired me to present dozens of conference papers and publish 13 books and more than 100 articles on topics such as waste reduction, science policy in northern Canada, occupational health issues such as asbestos, the privatisation of rail corridors and endangered species.

I also have been fortunate to teach at various post-secondary institutions such as York University and the University of Toronto between 1983 and 2018.

◀ **Top Tips**

1) Practicing environmental law these days requires some type of science background

   If you are interested in developing a successful environmental law practice, you will have to learn some basic science background in biology and chemistry, as well as understanding some engineering and math concepts. One senior environmental lawyer remarked to me in 2011, half joking, that environmental law today is 85% science and 15% law. There also are strong overlaps between environmental law, occupational health and safety regulations, public health, property law and numerous other areas of public and private law.

2) Research the university or colleges you are interested in attending and make sure they will help you develop the skills you need

   In response to student demand, many law schools and undergraduate programs are developing extremely innovative programs in environmental law and policy. Trent University, the university where I undertook my undergraduate studies in Canada, had programs in Canadian Studies, Indigenous/Aboriginal Studies, Environmental and Resource Sciences, Politics and Environmental History that were seen in the late 1970s, and remain today, as some of the most innovative in Canada. This provided me with some of the background I needed for my career.

3) Working on environmental law and policy requires interdisciplinary approaches

In order to provide sound advice, environmental lawyers must understand numerous interdisciplinary aspects of environmental problems including: the environmental impact of human-induced change at the local, regional and global levels; the role of technology and the policy process in determining how environmental problems evolve and are addressed; and various techniques for analyses to evaluate environmental tradeoffs such as cost-benefit analysis and environmental and social impact assessment. Lawyers must be able to integrate multiple disciplinary perspectives on environmental problems and understand various dimensions of environmental problems and environmental justice. For example, understanding how people interact with water on a human social scale in their daily lives (for drinking, cooking, farming, bathing, recreation, etc.) is an integral part of understanding and helping solve issues of clean water access in developed and developing nations.

4) Start writing for legal publications or magazines if you can

Many publications – both print and on-line – are seeking materials on environmental law issues or law reform. Take advantage of opportunities to write for publications as early as you can in your career to develop your skills.

5) Lecturing and Teaching law will help you sharpen your speaking and presentation skills and clarify your thinking

Teaching will allow you to share your insights about environmental law and policy and help you sharpen your speaking and presentation skills and clarify your thinking. Offer to do lectures and apply to teach at universities and colleges if you have the opportunity.

6) Many environmental interests and concerns such as protection of wildlife overlap with indigenous concerns

Environmental groups often have to work in partnership with indigenous peoples to ensure that development of forest, mineral, wildlife or energy resources takes place in an environmentally appropriate manner, minimizing impacts on wildlife habitat.

https://www.davidmcrobert.ca

https://www.linkedin.com/in/david-mcrobert-law/detail/recent-activity

https://www.yorku.academia.edu/DavidMcRobert

https://www.envstudiesyork.ca

Source: David McRobert, Environmental and IndigenousRights Lawyer and Author, Peterborough, Ontario, Canada. © John Wiley & Sons.

Personal Profile

Penny Simpson, Partner (Environment), Freeths LLP

Having trained as a solicitor in the City (Freshfields Bruckhaus Deringer's environmental law team) and with almost 25 years' experience in advising on environmental law in both private practice and in-house roles, Penny is a solid and experienced all-round environmental lawyer with particular expertise in 'natural environment' law.

Within the 'natural environment law' niche, she has built a strong, national reputation in advising a wide range of private and public sector clients. She advises on the regulation of protected habitats, protected species, water resources, air quality and environmental impact assessment.

Penny's clients are often commercial clients such as housing and commercial developers, environmental consultancies, mineral sector operators and water sector operators. Penny works with these clients to allow them to operate as effectively and efficiently as possible within the constraints imposed by natural environment law. Often, they need her help so as to deliver planning permissions and the other environmental consents (e.g. protected species licences, environmental permits, abstraction licences). However, Penny also advises local authorities and other Government bodies in the discharge of their environmental law functions, particularly their functions that relate to protection of the natural environment. And she also often advises community objectors to development and environmental pressure groups/organisations where they feel that environmental regulations are not being sufficiently adhered to by public bodies in making their decisions.

As well as advising on planning and consenting processes, Penny has significant experience in defending public and private sector clients when faced with environmental-related regulatory or criminal proceedings. These could be brought by the Environment Agency, Natural England, Natural Resources Wales, the Forestry Commission or the police (e.g. pollution offences, protected species offences, breaches of permits/licences).

More generally, she advises on the environmental aspects of property and company transactions where she negotiates to limit her client's environmental liabilities. She also has significant experience in health and safety law and frequently advises her clients on health and safety compliance.

Penny is listed as a "Leading Individual" in the Legal 500 (2020 edition) and a "Ranked Individual (Band 3)" in the Chambers (2020 edition).

Chambers: https://chambers.com/lawyer/penny-simpson-uk-1:128166

Legal 500: https://www.legal500.com/firms/1664-freeths-llp/5972-sheffield/england/lawyers/571020-penny-simpson/

◀ **Top Tips**

1) You don't have to study law at Uni if you want to become a lawyer – in fact it can sometimes be very helpful to study something else if you are going to be able to use it later in conjunction with a legal career: I was passionate about biology and

zoology when I was a youngster leaving school and wanted to study apes in the wild. I studied Biology at Uni and spent a year in West Africa tracking chimpanzees before I decided to come back and train as a lawyer, through the "conversion course" route. The Graduate Diploma in law is known as the conversion course and is designed for non-law graduates who want to progress a career in law. LawCareers.Net provides lots of helpful information and tips on how to progress a career in law (for both law and non-law graduates) including information on the future changes being made to the routes for becoming a lawyer. I don't regret for one moment any of that background because it set me up for my career as an interface between law and science

2) Put yourself out there – do something that other candidates have not done, so as to ensure you stand out: I remember that, the day before my interview with City law firm Freshfields for a training contract to become a solicitor, I phoned up Freshfields' Head of Environmental Law, got through and asked him all about his job and the work he did. I was then able to talk about that conversation in my interview and I am sure it was one of the reasons I got offered a training contract.

3) Follow your heart in your career progression: I made a career move early on in my legal career that everyone said was mad i.e. leaving a City law firm where I was well paid to go and work as the lawyer for an environmental pressure group. But the benefits of that move were in fact huge in terms of getting significant experience as the only lawyer in that organisation and in terms of getting into a new area of law (natural environment law) which I then used to develop my career further. So, don't be persuaded by what other people say and go with your heart – it'll work out in the end.

4) When you start work at an organisation, look around you at the senior people and ask yourself how have those people got to where they are? And try to start thinking that way from the start. For example, senior lawyers in law firms have to be great at winning clients, that is how they rise through the ranks. Be aware of that from the start and don't ever think 'that's not something I need to worry about at the moment' because the ones who succeed are the ones who realise early on that it is never too early to start doing this. Similarly, lawyers working in 'in house' (i.e. within a company organisation or a charity) need to be really good a spinning lots of plates at once, often leaving the detail to external law firm lawyers who they instruct to assist them.

5) Be passionate about what you do – that way you will attract people towards you, and you will find that you become a natural and effective communicator, an essential skill in many jobs

https://www.freeths.co.uk/people/penny-simpson

Source: Penny Simpson, Partner (Environment), Freeths LLP. © John Wiley & Sons.

Company Profile

Freeths LLP's Environmental Law Team

Freeths LLP's Environmental Law team is made up of two partners who lead the team, solicitors, legal assistants and trainees.

Freeths LLP's Environment law practice has significant expertise and experience across a full range of environmental law work including:

Advising on environmental permitting i.e. where consents/permits are needed to carry out operations which might negatively affect the environment

Advising on the complex issues of waste regulation and compliance

Advising on law that protects the natural environment e.g. protected species and habitats, water resources and environmental impact assessment

Advising clients on how to defend themselves from criminal and regulatory action where allegations are made that have breached controls

Advising on noise and odour nuisances and how to stop/control them or how to defend a client if they are producing the noise

Advising on disputes between landowners relating to pollution or other environmental issues

Advising on challenging public body/Government decisions which have not complied with environmental law standards

Providing day long training courses on environmental law issues

LinkedIn: https://www.linkedin.com/company/freeths-llp

Twitter: https://twitter.com/freeths

Sources: Adapted from LinkedIn, Twitter, LinkedIn: https://www.linkedin.com/company/freeths-llp Twitter: https://twitter.com/freeths. © John Wiley & Sons.

Personal Profile

Fabiano de Andrade Correa, International Consultant on Sustainable Development Law and Policy, Brazil

I grew up in close contact with nature in the South of Brazil and was lucky to be raised in a family that loved travelling and the outdoors. My love for nature and a curiosity to travel the globe and discover different cultures and ways of living made me interested in pursuing a diplomatic career. I decided to study Law with this idea in mind (in Brazil you need a university degree in order to take the public examination to the Diplomatic Academy).

Over the years in Law School, in particular taking the (back then, in many places still to this day) optional class on environmental law, I became fascinated with the concept of sustainable development and decided to deepen my studies around it. I ended up practicing law for a couple of years, but ultimately decided to go back to my original idea. The way I found to do this was to go abroad and do a Masters in International Relations and Diplomacy, with a specialisation in development cooperation. It was an incredible experience, which made me decide to become a specialist in law and sustainable development rather than going forward with a general diplomatic role. I got a scholarship to do a PhD in International Law, which I leveraged as an opportunity to both deepen my theoretical knowledge, but also to broaden my network, participating in as many conferences, events and networks as I could. This helped me to eventually find a job as international legal consultant with an international organisation during the last year of my doctoral studies in 2012 – and I have been working in this field since then.

It has been a fascinating experience, having worked in countries across 4 continents on several topics like climate change, biodiversity, human rights, land and tenure, trade and investment, having sustainable development as a guiding theme. My role includes legal assistance to governments in implementing international instruments, organising awareness raising events and capacity development, as well as research and publications. I am also a guest lecturer, and volunteer in different think tanks and networks related to these topics.

I used to believe there would be more work in this area with national governments and international organisations, but over the last years I have been noting an increasing uptake of sustainability also within civil society organisations and the private sector, especially during the past two years.

I see this as an expanding field, in which lawyers and legal advisors have an important role to play, from advising international negotiations; working to support implementation and enforcement of international agreements at different levels; advising the drafting and implementation of national legislation and policies; working on the litigation side – for instance, public interest litigation, climate change cases; and advising clients and corporations on the business case for embracing sustainability practices and solutions, which are badly needed to advance these agendas.

◀ Top Tips

1) Find a topic that you are really passionate about: this might sound cliched, but working on something you are really interested in makes it much easier to do what it takes to become really good at it.
2) Pursue the best educational path that you can: it still makes a big difference on your CV. I am often asked if I recommend doing a PhD, for instance, and my advice is usually yes, with the caveat that you should understand your goal with it, and orient it accordingly – e.g. if you want to do a career in academia, it is fine to focus on a very theoretical topic; whereas if you'd like to move to practice afterwards, a more practical approach might make it easier to apply your knowledge afterwards. Lack of resources is not an excuse, as there are many scholarships available, it is time consuming but worth checking.
3) Pursue 'relevant experiences' as early as you can, and build 'skills' – these can range from public speaking, management, analytical thinking, text editing, social media – in addition to gaining theoretical knowledge. The world is full of problems and needs problem solvers.

4) Expand your networks, find people/organisations that are good in what you want to do and try to follow them at conferences, events, social media, even internships/volunteering, which are a great entry point to gain experience. Through a broad network you are not only constantly updated on current topics but also on potential professional opportunities.

5) Understand that a professional in this field has a double role: that of working to support existing needs, but also raising awareness of the increasing environmental challenges we are facing and opening new work opportunities as you do it.

www.fabianodeandrade.com

https://www.linkedin.com/in/fac82

Source: Fabiano de Andrade Correa, International Consultant on Sustainable Development Law and Policy, Brazil. © John Wiley & Sons.

4.9   External Resources

UK

The UK Environmental Law Association
Website section aimed at students
www.ukela.org
www.environmentlaw.org.uk – an easy to read guide to environmental law

Environmental Law Foundation
www.elflaw.org

Friends of the Earth, legal internships
https://www.jobs.friendsoftheearth.uk/vacancies

Government Legal Department
https://www.gov.uk/government/organisations/government-legal-department/about/recruitment

Lawcareers.net

Student guides to the legal profession have useful information on working in environmental law
www.lawcareers.net

Chambers Student Guide (23rd Edition was published in 2019)
www.chambersstudent.co.uk

Target Law
https://www.targetjobs.co.uk/career-sectors/law-solicitors

Client Earth
https://www.clientearth.org/how-to-become-an-environmental-lawyer-in-the-uk

Europe

European Union Network for the Implementation and Enforcement of Environmental
Law (IMPEL)
www.impel.eu

Euro LegalJobs
https://www.eurolegaljobs.com/job_search/category/environment_and_energy_law

Greenpeace International, internships
www.workfor.greenpeace.org/interns

Asia

LAWASIA
www.lawasia.asn.au
www.lawasiafiji2020.com/conference-programme

Asia Pacific Journal of Environmental Law
https://www.elgaronline.com/view/journals/apjel/apjel-overview.xml

Chinese Journal of Environmental Law
https://www.brill.com/view/journals/cjel/cjel-overview.xml

Africa

www.africa-legal.com/jobs
https://www.gostudy.net/occupation/environmental-lawyer

Law Society of South Africa
www.lssa.org.za

Environmental Law Association South Africa
www.elasa.co.za

North America

Environmental Law Institute
www.eli.org

Law Crossing
www.lawcrossing.com

Eco Justice
www.ecojustice.ca

Harvard Law School – A Trail Guide to Careers in Environmental Law
https://www.hls.harvard.edu/content/uploads/2008/07/full-working-draft.pdf

Yale Law School – Environmental Law
https://www.law.yale.edu/sites/default/files/area/department/cdo/document/cdo_
environmental_law_public.pdf

South America

National Law Review – Latin American Environmental Regulatory Tracker
https://www.natlawreview.com/article/latin-american-environmental-regulatory-tracker-
    december-16-2018-january-15-2019

ORyan R., Ibarra C. (2016) Environmental Policy in Latin America. In: Farazmand A. (eds)
    Global Encyclopedia of Public Administration, Public Policy and Governance. Springer, Cham.
    www.doi.org/10.1007/978-3-319-31816-5_2670-1

Oceania

National Environmental Law Association (NELA)
www.nela.org.au

Environmental Justice Australia
www.envirojustice.org.au/who-we-are

Environmental Law Australia
www.envlaw.com.au
www.envlaw.com.au/links

New Zealand Centre for Environmental Law
https://www.auckland.ac.nz/en/law/our-research/research-institutes-centres/new-
    zealand-centre-for-environmental-law.html

Australia public jobs
www.apsjobs.gov.au

Global

The Legal 500
www.legal500.com

UN Jobs
www.unjobs.org/themes/environmental-law

Client Earth
www.clientearth.org

**FOR UPDATED AND ADDITIONAL RESOURCES, INCLUDING EXTRA CHAPTERS, GO TO
WWW.ENV.CAREERS**

References

All About Law. (2015). Environmental Law. www.allaboutlaw.co.uk/stage/areas-of-law/
    environmental-law (accessed 24 January 2016).

Australian Government Department of the Environment. (2015). Environment Protection and
Biodiversity Conservation Act 1999. www.environment.gov.au/about-us/legislation
(accessed 27 January 2016).

Beveridge & Diamond, P.C. (2016). Latin American Practice http://www.bdlaw.com/
practices-122.html (accessed 25 January 2016).

Boer, B. (ed.) (2015). Environmental Law and Human Rights in the Asia-Pacific (July 8, 2014).
In: *Environmental Law Dimensions of Human Rights*. UK: Oxford University Press Sydney
Law School Research Paper No. 14/62. www.ssrn.com/abstract=2463525.

Boer, B. (1999). The rise of environmental law in the asian region. *University of Richmond Law
Review* 32: 1503. https://www.scholarship.richmond.edu/cgi/viewcontent.cgi?article=2320&
context=lawreview.

Centre for Environmental Rights. (2016). The Centre for Environmental Rights is a non-profit
company and law clinic based in Cape Town, South Africa

Centre for Environmental Rights. (2021). www.cer.org.za/about/overview. (accessed 5
April 2021).

Environmentalscience.org.(2021). 'What is an Environmental Lawyer?' https://www.
environmentalscience.org/career/environmental-lawyer. (accessed 5 April 2021.

EUR-Lex. (2015). Environment and climate change https://www.eur-lex.europa.eu/summary/
chapter/environment.html. (accessed 23 January 2016).

Goldman, L. (2013). Protecting Villagers From the Health Effects of Gold Mining. https://
www.eli.org/sites/default/files/docs/research-action_lisa.pdf (accessed 20 January 2016).

Harashima, Y. (2000). Environmental governance in selected asian developing countries.
International Review for Environmental Strategies 1 (1): 193–207. 2000.

Harvard Law School (2013). *A Trail Guide to Careers in Environmental Law' Bernard Koteen
Office of Public Interest Advising*. Harvard Law School.

Kameri-Mbote, P., Paterson, A., Ruppel, O., Orubebe, B.B., Kam Yogo, E.D., (Eds.) (2019) Law |
Environment | Africa, 21–30 Publication of the 5th Symposium | 4th Scientific Conference |
2018 of the Association of Environmental Law Lecturers from African Universities in
cooperation with the Climate Policy and Energy Security Programme for Sub-Saharan
Africa of the Konrad-Adenauer-Stiftung and UN Environment

McRobert, D.S. and Ruby, G. (2008). Law and Sustainability: The Canadian Case. In:
Introduction to Sustainable Development (eds. D.V.J. Bell and Y.A. Cheung) Encyclopedia of
Life Support Systems (EOLSS), Developed under the Auspices of the UNESCO. Oxford, UK,
(http://www.eolss.net) Available from:: Eolss Publishers (accessed 29 September
2020).https://www.researchgate.net/
publication/282663187_LAW_AND_SUSTAINABILITY_THE_CANADIAN_CASE

Raine, A. and Pluchon, E. (2019). UN Environment—Advancing the Environmental Rule of
Law in the Asia Pacific. *Chinese Journal of Environmental Law* 3 (1): 117–126. https://doi.
org/10.1163/24686042-12340037.

Saxe, D. (2013). So you want to be an environmental lawyer. . .. www.siskinds.com/
environmental-lawyer (accessed 5 April 2021).

Tan, A. (2004). Environmental laws and institutions in southeast asia: a review of. *Singapore
Yearbook of International Law (SYBIL)* VIII (1): 177–192.

The Legal 500 'Environment' (2020). www.legal500.com/c/london/real-estate/environment.
(accessed 10 September 2020).

The Legal 500 'United States' (2020). www.legal500.com/c/united-states (accessed 10 September 2020).

The University of Law (2013). Environmental Law https://www.law.ac.uk/employability/legal-practice-areas/environmental-law (accessed 28 January 2016).

United Nations (1948) 'The Foundation of International Human Rights Law'. https://www.un.org/en/about-us/udhr/foundation-of-international-human-rights-law#:~:text=The%20Universal%20Declaration%20of%20Human,binding%20international%20human%20rights%20treaties (accessed 5 April 2021).

United Nations Environment Program (UNEP) (2019). Air Pollution in Asia and the Pacific: Science-based Solutions.

UNEP (2016). The Global Environmental Outlook GEO-6 – Regional Assessment for Africa. https://www.unenvironment.org/resources/global-environment-outlook-6-regional-assessments (accessed 6 April 2021).

Zhao, J. (2019). *Environmental Regulation: Lessons for Developing Economies in Asia. ADBI Working Paper 980*. Tokyo: Asian Development Bank Institute https://www.adb.org/sites/default/files/publication/514476/adbi-wp980.pdf.

Brundtland, G.H. (1987). *Our Common Future: Report of the World Commission on Environment and Development*. Geneva, UN-Document A/42/427.

United Nations (2015). The Paris Agreement https://unfccc.int/sites/default/files/english_paris_agreement.pdf (accessed 5 April 2021).

Australian Government (1999). Environment protection and biodiversity conservation act 1999. https://www.legislation.gov.au/Details/C2016C00777 (accessed 5 April 2021).

ClientEarth (2019). How to become an environmental lawyer in the UK. https://www.clientearth.org/latest/latest-updates/stories/how-to-become-an-environmental-lawyer-in-the-uk/ (accessed 5 April 2021).

5

Environmental Policy, Legislation and Regulation

5.1 Sector Outline

Environmental Policy is a wide-ranging field that encompasses several sectors – Government departments, NGOs, consultants and others. Historically, Government departments developed most areas of national policy using their own staff, but this has altered in recent decades so that Government endorsed working groups of policy professionals develop policy in a more inclusive way. These policy professionals can come from NGOs, trade bodies, charities and consultancies. However, this often means that Government departments have become denuded in terms of their policy making ability.

There are also groups called Think Tanks, which are expert groups providing evidence, generally on economic and political issues.

According to the LSE in its online Careers Resource (2020):

> What think tanks do depends on their size and funding, but principally their aim is to publish and influence public policy debate. Most think tanks are non-profit organisations and may be based in the charity sector. Others are funded by particular advocacy groups, the voluntary sector, government, businesses, generate revenue from consulting or research or combination thereof. Although these policy institutes may have political bias they are usually independent of political parties and government. Think tanks conduct research into a range of issue oriented topics as well as broader business, political and economic areas including different aspects of social policy, political strategy and reform, the environment, security and the military, technology and economics using a variety of social science research methods. Think tanks are no longer a North American and Western European phenomena and indeed the past 10 years has witnessed the impact of globalization on the think tank movement. This is most evident in regions such as Africa, Eastern Europe, Central Asia, and parts of Southeast Asia. Here there has been a concerted effort by the international community to support the creation of independent public policy research organizations resulting in collaboration across geographical regions.
>
> The think tank sector employs relatively few people. The smallest institute may have a core staff of 3 or 4 and employ extra staff when needed. The largest employ around 50.

5.2   Issues and Trends

A word that is often used in terms of policy development is 'lobbying', and this is still relevant in terms of policy professionals needing to be persuasive in terms of getting their views across, but policy development has tended to be less combative in recent years worldwide. Roles tend to lean more towards campaigning for NGOs whose main objective is to change policy to work towards their organizational aims.

There are policy professionals working for trade bodies and large conglomerates who often work against more effective environmental policy development, and there is strong traditional lobbying from this sector. However, there is a growth in philanthropic green policy support that counters the negative approaches by some.

In recent years, there has been a significant growth in consultancies actively being involved in policy development work and they have been developing in-house policy teams.

5.3   Key Organisations and Employers

Key organisations and employers include government departments, NGOs, trade bodies, consultancies, think tanks and large corporations. Government departments are the largest employers as policy specialists are active in developing, discussing, negotiating and implementing new government policy.

5.4   Careers in the Sector

The policy sector has a varied composition of people from backgrounds in consultancies, trade bodies, government and the private sector. Within the sector, key people become 'known' for policy areas such as water, waste, resources and planning and work on an ongoing set of regulations and policies relating to that field. They then become quite specialised in specific sectors. The largest sector is within government, so that is a key focus for initial roles – to secure a position in a government department. This can then be a springboard to working within an NGO, think tank or trade body or many employees stay within government as it offers an opportunity to develop and implement policy from within.

There are several websites and newsletters where the key policy jobs and issues are highlighted. As environmental policy is a fairly small sector, you need to know where to look! Internet searches aren't very helpful at finding the key jobs. Many organisations with policy jobs cannot afford to be listed on lots of big and expensive recruitment sites, making job searches much more complicated.

5.5   Job Titles in the Sector

Key job titles: Policy officer, policy manager, regulatory policy manager, regulatory advisor, policy analyst, government affairs advisor, campaign officer.

5.6   Educational Requirements

There are more environmental policy courses and degrees than ever before, but the majority of people involved in environmental policy come through a non-direct education. Many policy professionals have Geography degrees or environmental degrees. Most policy staff are educated to degree level or above, but this is not a prerequisite for working in the sector.

5.7   Personal Attributes and Skill Sets

Environmental policy work needs a range of specific skills, including:

- Well established communication skills
- Quick assimilation of information
- Negotiation and persuasion skills
- Proof reading and document drafting
- Working in a group environment to positively negotiate and develop policy
- In depth sector knowledge
- Network development and use

Jobs tend to be more office based, which may be an issue for environmental professionals who love being outdoors.

Environmental policy professionals are a mixture of specialists and generalists – some focus on one area of work such as water policy and act as a source of expertise and others cover a wide range of policy areas and call upon experts for specific areas of coverage.

5.8   Career Paths and Case Studies

◀ **Top Tips**

Dr Maria D Carvalho, Policy analyst at the Grantham Research Institute on Climate Change and the Environment, UK

1) Realise that both academia, firms and civil society are integrating environmental and social sustainability as a key issue to address. Therefore, the breadth and scope of environmental-related careers is increasing, and it can be a core function of your future job, or increasingly something that relates to your job. Therefore, don't keep a narrow perspective on what it means to have an environmental career, and realise that everything from banks to mining will need to have people with environmental expertise. The good news is that expands the kind of job opportunities you can pursue.
2) Figure out what kind of industry and environmental issues you are interested in learning about. See if there are any think tank reports that can help you understand the key challenges and opportunities to enabling environmental sustainability. It is necessary to do this kind of research throughout your career in order to: (i) continuously reinforce your passion and curiosity, which are your guiding lights in determining which pathway

you want to take in your career, and necessary in challenging times to remind you why you are pursuing a career to protect the environment. (ii) help you identify key organisations you would want to work with and (iii) identify key individuals you would want to speak to, or at the very least, see things of career trajectory they did to have the career they got into (e.g. using LinkedIn).

3) Go to conferences and talks on issues you are interested in. If it is expensive fee, see if you can volunteer with the organising. Networking is still a key way for people to meet people to get internships and jobs, and getting access to conferences is a good way to understand the latest issues and the people who are discussing it.

4) If you are still in school, do internships to figure out what kind of environmental jobs and organisations that you would like to work for. Also volunteer in environmental related extracurricular activities (there is a student group in UK called the Dirty Weekenders) and they rehabilitate environmental spaces). This is important for you in reminding yourself what you are trying to protect, as well as is a good signal in your CV of your commitment.

Always let passion and curiosity be your guiding lights – it is hard to effectively map out your career a priori because you learn a lot about yourself along the way. But passion and curiosity help guide the direction you want to go. In letting passion and curiosity guide you, train yourself to think of constructive solutions to environmental challenges. Regardless of which type of environmental career you choose (or any career), people are attracted to those who provide solutions, or are willing to undertake a solutions building process. And we need effective solutions to protect the environment.

Source: Dr Maria D Carvalho, Policy analyst at the Grantham Research Institute on Climate Change and the Environment, UK. © John Wiley & Sons.

◀ Top Tips

Justine Saunders, Environmental Policy Advisor, Singapore

My top 5 tips for the environmental sector (probably not too different from other sectors):

1) Always follow your passion in life. It will be a much more enriching and rewarding experience. Environmental sciences cover such a broad range of skills that there is a job for everyone whether you're a designer (to help communicate scientific results), software programmer or a good project manager. Think about your strengths and your passions and try to combine them.

2) Given such diversity and the rapid changes in the world, be flexible. Keep an open mind. My career has evolved in ways that I never would have predicted. Look for opportunities to pick up relevant new skills and take them – at any age – whether you're 20 or 60. We will likely all be living to 100+ by 2030 so it's a long road ahead.

3) Stay balanced. Every scientist I know has another talent whether it's a creative musical or artistic talent or sport. You might not think that it's relevant as career advice but the more balanced you are, the happier you will be and the more that you will be able to bring to your career.

4) Stay connected. Don't forget why you're doing this. Science can sometimes be so academic and conceptual that you can lose touch with the real world. Keep in touch with society and humanity. What are the real issues? What do we need to survive and be happy? How can your work contribute to that? Talk to people from school children to your taxi driver to government decision-makers. I've learnt more from them than many lectures.

5) Don't be afraid. Don't be afraid to ask questions or to make mistakes. One of the biggest criticisms at the moment of science is that we don't report our mistakes or failures or more specifically failures to disprove a null hypothesis with the result that our colleagues head down the same wrong path too. Despite the competitive structures that fund research, we're in this together. With us all racing to solve critical issues in environmental science, from climate change to plastic pollution, there is no room for ego.

Source: Justine Saunders, Environmental Policy Advisor , Singapore. © John Wiley & Sons.

◀ Top Tips

Martin Baxter, Chief Policy Advisor, IEMA, UK

1) Do voluntary experience or an internship within the industry.
2) Get an environment and/or sustainability-related degree that is approved by a professional body so you can get graduate member status ready for when you start looking for work.
3) Seek mentoring from someone already doing the job you want.
4) Get the right vocational qualifications like the IEMA Foundation Certificate.
5) Join a Professional Body which will guide, support and celebrate your work and development throughout your career.

Source: Martin Baxter, Chief Policy Advisor, IEMA , UK. © John Wiley & Sons.

5.9 External Resources

UK

LSE, Think tank careers globally by region
www.info.lse.ac.uk/current-students/careers/resources/type-of-organisation/think-tanks-by-region
www.lse.ac.uk/intranet/CareersAndVacancies/careersService/EmploymentSectors/Public SectorPoliticsPolicy/ThinkTanks/Home.aspx

The Guardian's list of Think Tanks
www.theguardian.com/politics/2013/sep/30/list-thinktanks-uk

Oxford University list of Think Tanks – also international think tanks listed
www.careers.ox.ac.uk/think-tanks

PubAffairs website
www.publicaffairsnetworking.com/public-affairs-jobs.php

The specialist recruiter Michael Page is a good source for UK Policy jobs
www.michaelpage.co.uk
www.michaelpage.co.uk/expertise/policy

Charity Job website
www.charityjob.co.uk

Guardian jobs website
www.jobs.theguardian.com/jobs/policy

Third Sector Jobs
www.jobs.thirdsector.co.uk/jobs/policy-and-research/

Environment Job
www.environmentjob.co.uk/jobs

Greenjobs
www.greenjobs.co.uk/

Europe

Politjobs
www.politjobs.eu/

Euro Brussels
www.eurobrussels.com/

EU Internships (stagiaires) – placements with EU departments
www.europa.eu/epso/working/training_en.htm
www.epso.europa.eu/job-opportunities/traineeships_en

Asia

ILO 'Green Jobs in Asia and the Pacific'
www.ilo.org/asia/areas/green-jobs/lang--en/index.htm

UNDP
www.asia-pacific.undp.org/content/rbap/en/home/operations/jobs.html

UN Jobs
www.unjobs.org/themes/southeast-asia

Africa

African Institute for Development Policy (AFIDEP)
www.afidep.org/jobs

Career Jet
www.careerjet.co.za/public-policy-jobs.html

North America

Green Jobs
www.greenjobs.net

Environmentalscience.org
www.environmentalscience.org/careers/environmental-policy-and-planning

Ecojobs
www.ecojobs.com/environmental-policy-and-regulation-jobs.htm

University of Maryland, Internships
www.ensp.umd.edu/internships/internships-policy

Yale School of the Environment – resources section
www.environment.yale.edu/careers/resources/jobs-sites

International Relations UDU
www.internationalrelationsedu.org/foreign-affairs-analyst

Global Policy Lab
www.globalpolicy.science/jobs

South America

UN Jobs Latin America
www.unjobs.org/themes/latin-america

Economic Commission for Latin America and the Caribbean
www.intjobs.com/jobs_at/un_eclac_economic_commission_for_latin_america_
   and_the_caribbean/5279

Latin American Network Information Center (LANIC)
www.lanic.utexas.edu/enlace/resources

Oceania

NZ Government jobs
www.jobs.govt.nz

Australia Government/public service policy jobs:
www.apsjobs.gov.au

Ethical Jobs
www.ethicaljobs.com.au/policy-research-jobs

Global

SEI global internships and jobs
www.sei-international.org/working-for-sei

UN Jobs
www.unjobs.org/themes/public-policy

International Affairs Jobs
www.intjobs.com/

Environmentalscience.org 'Environmental Policy & Planning Careers'
www.environmentalscience.org/careers/environmental-policy-and-planning

Global Jobs
www.globaljobs.org/

Bond – international development jobs
www.bond.org.uk/jobs

Institute for Transportation and Development Policy
www.itdp.org/careers

Impact Pool
www.impactpool.org/search

Global Policy Forum
www.globalpolicy.org/internships/general-information-on-internships.html

GPPI
www.gppi.net/about/jobs-internships/

UNDP Jobs
www.undp.org/content/undp/en/home/jobs.html

National Institute for Research Advancement – links to research institutes in more than 50 countries
www.english.nira.or.jp/
www.english.nira.or.jp/directory/

Policy Library
www.policylibrary.com

FOR UPDATED AND ADDITIONAL RESOURCES, INCLUDING EXTRA CHAPTERS, GO TO WWW.ENV.CAREERS

Reference

London School of Economics and Political Science (LSE),(2020) LSE Careers, Think Tanks https://info.lse.ac.uk/current-students/careers/information-and-resources/type-of-organisation/think-tanks (accessed 6 April 2021).

6

Conservation and Ecology

6.1   Sector Outline

Working in conservation is a common aim for someone keen to work in the environment sector and this career choice feels like a direct response to wanting to make a difference.

There is a lot of crossover between this chapter and those on consultancy and the Emerging Sector section on NGOs, as significant conservation work is delivered through NGOs, and consultancies have growing employment for conservation and ecology professionals.

As policy and legislation grows in the sector, so do the number of jobs.

Lucas, Gora and Alonso (2017) note:

> Conservation is now a primary goal of most national governments as indicated by the International Union for Conservation of Nature's implementation of the strategic plan for biodiversity conservation (United Nations Convention on Biological Diversity 2011) and with 187 countries committing to decreasing carbon emissions at the 2015 Intergovernmental Panel on Climate Change Conference. With these extensive commitments to conservation worldwide, the need for well-trained scientists and managers in the field of conservation is at its highest.
>
> Due to the global nature of conservation, positions in conservation management and research are based in countries with different levels of economic development (hereafter development). Level of development often has a large impact on a country's strategy and ability to address conservation challenges.
>
> Countries with developing economies often struggle with a perceived trade-off between economic growth and conservation (Czech 2008) and may not have sufficient funds to meet conservation needs.
>
> Advanced economies may have more funds available, and organizations frequently collaborate with the private sector to create positive conservation outcomes (Robinson 2012).

Global Environmental Careers: The Worldwide Green Jobs Resource, First Edition. Justin Taberham.
© 2022 John Wiley & Sons Ltd. Published 2022 by John Wiley & Sons Ltd.

6.2   Issues and Trends

There is a global trend towards consultancies taking on monitoring and research work which traditionally would have been carried out by government agencies. There is also a growth in the number of smaller companies or individual consultants who develop targeted focus in a specific area, such as bats, insects, reptiles or plant management. Often, smaller companies are being taken over by the larger consultancies but there is a growing market for specialists who are contracted by governments, consultancies, developers and others to carry out specific tasks.

The growing standardisation of monitoring techniques and growth in global knowledge base are enabling more professionals to work globally in this sector rather than be based locally or nationally.

6.3   Key Organisations and Employers

The Conservation Careers website (2018) listed its top employers in the sector, based on their job listings.

World Wide Fund for Nature (WWF)

The mission of WWF is to stop the degradation of the planet's natural environment and to build a future in which people live in harmony with nature, by: conserving the world's biological diversity, ensuring that the use of renewable natural resources is sustainable and promoting the reduction of pollution and wasteful consumption.

Royal Society for the Protection of Birds (RSPB)

The RSPB is the United Kingdom's largest nature conservation organisation and charity, inspiring everyone to give nature a home. Together with our partners, they protect threatened birds and wildlife so that towns, coast and countryside will once again teem with life.

The Nature Conservancy (TNC)

The mission of The Nature Conservancy as a conservation organisation is to conserve the lands and waters on which all life depends. Its vision is a world where the diversity of life thrives, and people act to conserve nature for its own sake and its ability to fulfil their needs and enrich their lives.

New Zealand Department of Conservation (NZ DOC)

The Department of Conservation is the public service department of New Zealand charged with the conservation of New Zealand's natural and historical heritage. An advisory body, the New Zealand Conservation Authority is provided to advise DOC and its ministers.

The Wildlife Trusts

The Wildlife Trusts, the trading name of the Royal Society of Wildlife Trusts, is an organisation made up of 46 local Wildlife Trusts (county-based conservation organisations) in the United Kingdom, the Isle of Man and Alderney. The Wildlife Trusts, between them, look after around 2300 nature reserves covering more than 98 000 hectares.

The National Trust

The National Trust, formally the National Trust for Places of Historic Interest or Natural Beauty, is a conservation organisation in England, Wales and Northern Ireland, and the largest membership organisation in the United Kingdom.

The International Union for Conservation of Nature (IUCN)

The IUCN is an international conservation organisation working in the field of nature conservation and sustainable use of natural resources. It is involved in data gathering and analysis, research, field projects, advocacy and education.

Conservation International (CI)

Conservation International is an American nonprofit environmental organisation headquartered in Arlington, Virginia. Its goal is to protect nature as a source of food, fresh water, livelihoods and a stable climate. CI's work focuses on science, policy, and partnership with businesses and communities.

BirdLife International

BirdLife International is a global partnership of conservation organisations that strives to conserve birds, their habitats and global biodiversity, working with people towards sustainability in the use of natural resources.

Pew Charitable Trusts

The Pew Charitable Trusts is an independent nonprofit, non-governmental organisation, founded in 1948. With over US $6 billion in assets, it uses evidence-based, non-partisan analysis to solve today's challenges. Pew's work helps preserve wilderness, restore biodiversity and increase understanding of ocean ecology.

British Antarctic Survey (BAS)

BAS delivers and enables world-leading interdisciplinary research in the Polar Regions. Its skilled science and support staff based in Cambridge, Antarctica and the Arctic, work together to deliver research that uses the Polar Regions to advance our understanding of Earth as a sustainable planet.

Blue Ventures

Blue Ventures develops transformative approaches for catalysing and sustaining locally led marine conservation. It works in places where the ocean is vital to local cultures and economies and is committed to protecting marine biodiversity in ways that benefit coastal people.

Zoological Society of London (ZSL)

ZSL is an international conservation and scientific charity based in the United Kingdom. Founded in 1826, ZSL is an international scientific, conservation and educational charity whose mission is to promote and achieve the worldwide conservation of animals and their habitats. ZSL's mission is realised through its groundbreaking science, active conservation projects in more than 50 countries and two zoos, ZSL London Zoo and ZSL Whipsnade Zoo.

The National Audubon Society (Audubon)

Audubon is a nonprofit conservation organisation dedicated to conservation. Located in the United States and incorporated in 1905, Audubon is one of the oldest of such organisations in the world and uses science, education and grassroots advocacy to advance its conservation mission.

Student Conservation Association (SCA)

SCA's mission is to build the next generation of conservation leaders and inspire lifelong stewardship of the environment and communities by engaging young people in hands-on service to the land.

Wildlife Conservation Society (WCS)

Wildlife Conservation Society was founded in 1895 as the New York Zoological Society and currently works to conserve more than two million square miles of wild places around the world.

Fauna & Flora International (FFI)

Established over a century ago, FFI was the world's first international wildlife conservation organisation. Its mission is to conserve threatened species and ecosystems worldwide, choosing solutions that are sustainable, based on sound science, and which take into account human needs.

6.4   Careers in the Sector

This is a sector where competition for jobs is intense. Salaries are also generally low, so there needs to be an understanding of the challenge you face with developing a career in conservation.

Volunteering is a key issue in working in conservation and ecology as well as in the wide environment sector. Volunteering is a way to develop skills and experience and you can volunteer as a young person, so you can get an early start to developing your green skills.

There is a section on volunteering in the book Introduction.

6.5   Job Titles in the Sector

Ecologist, conservation officer, conservation manager, wildlife biologist, biologist, researcher, programme manager, project manager, environment officer, specialist, adviser, conservation analyst, conservation coordinator, field manager, campaigner.

6.6   Educational Requirements

There is a considerable debate as to whether a higher-level degree is needed in the sector. This is a very competitive sector, but employers will often be seeking experience rather than just qualifications. However, some roles directly ask for master's level degrees.

There are many routes into the sector which don't need a degree, which often focus on site and habitat management work.

6.7   Personal Attributes and Skill Sets

In the 2017 Conservation Biology paper 'A view of the global conservation job market and how to succeed in it', authors Lucas, Gora and Alonso addressed some challenges the sector faces.

> Despite the increased demand for conservation scientists, conservation-based training has lagged, with limited information describing how to receive appropriate training. If successful conservation requires a range of skill sets, then it is imperative that well-informed and focused training begins early to ensure that conservation scientists are equipped to combat Earth's conservation needs and find employment.
>
> These fundamental differences in resources and historic presence of conservation likely necessitate different types of skills and training to succeed as a conservationist.
>
> Properly educating students on how to prepare for international positions and collaborations will increase both their future prospects and the quality of conservation efforts worldwide.
>
> Mastering both general and specific skills related to scientific practice is the primary goal of graduate school. Similarly, mastering an analytical tool, such as GIS, is one way to make oneself marketable across sectors.
>
> For students interested in continuing in academia, teaching experience is one of the top skills to develop throughout their education, along with publishing in peer-reviewed journals and acquiring funding.

If students plan to work outside academia, we recommend they focus on gaining experience in business and project management and developing interdisciplinary skills.

Some schools in the United States are beginning to incorporate paid internships with external organizations as a way to supplement student income. . .These programs provide meaningful conservation-based training that can lead to important networking benefits.

One of the more striking differences between the job markets of developed and developing economies was how important international, cultural, and language based experience was for working in countries with developing economies.

For students interested in working in a specific country long term, it would be beneficial for them to conduct their dissertation research in their country of interest. We also found that interpersonal skills were more frequently required in jobs in developing countries.

Therefore, encouraging international collaborations early on in an individual's career may be one of the best ways to prepare her or him for the future international job market. Developed nations frequently emphasized skills that are accessible in most graduate programs (e.g., technical and analytical skills, publication history, and written communication).

To be successful in today's job market, students must take command of their education early on. This may require looking outside of their departments and institution to diversify their training. We recommend students speak with conservation professionals early in their education to identify the sector that best fits their goals.

Skill sets and knowledge which are of help in the sector include:

- Normally, a degree in a relevant (normally natural sciences) subject.
- Data analysis and interpretation.
- Specific knowledge, certification and licensing for specific species, e.g. bats, mammals, birds, protected species.
- Previous relevant experience.
- Project management.
- Teamwork.
- Report writing and presentation skills.
- Committee reporting.
- Communications skills.
- IT skills, often in statistical analysis and GIS/mapping.
- Management skills.
- Commitment, passion and self-motivation.

Training

Steph into Nature – 11 Sites for Free Online Wildlife Conservation Courses
www.stephintonature.com/blog/free-online-wildlife-conservation-courses

Conservation Jobs
www.conservationjobs.co.uk/courses

Listing of online courses by The Wildlife Society
www.wildlife.org/next-generation/career-development/online-courses

Wildwood Ecology Training Academy
www.facebook.com/groups/197991751531444/permalink/240757190588233

Conservation Leadership Program
www.conservationleadershipprogramme.org

Conservation Training
www.conservationtraining.org

Emerging Wildlife Conservation Leaders
www.wildlifeleaders.org/

National Science Foundation Integrate Graduate Education and Research Traineeship
www.nsf.gov/crssprgm/igert/intro.jsp

Wildlifetek
www.wildlifetek.com

University of London Short Course Conservation and Society
www.soas.ac.uk/courseunits

Smithsonian-Mason School for Conservation Professional Training Programs
www.smconservation.gmu.edu

Wildlife Conservation Society Biodiversity Conservation
Professional Certificate Program
www.wildlife.org/learn/professional-developmentcertification/certification-programs

Program on African Protected areas & Conservation (PAPACO)
A series of free online courses
www.papaco.org/mooc-on-ecological-monitoring

6.8   Career Paths and Case Studies

Building knowledge and networks is important and this can be by attending courses and talks. Professional and trade bodies often have student and local groups. In the sector, a high proportion of jobs are not advertised widely, so keeping an eye out through sector networks is very important. Competition for all roles is very high.

Case studies

◀ **Top Tips**

Claire Wansbury, Associate Director of Ecology, Atkins Ltd, UK

1) Get experience as well as qualifications. It will help your job prospects, but also it will help you think how you might want to focus your career. Taking consultancy as an example – while it is great to be out in the field a lot, you also need to learn reporting skills, so look for opportunities to develop these alongside field ecological experience in your early career.

2) Make contacts, and don't be afraid of taking chances. One very fortunate step was my first paid job at the Countryside Council for Wales. I had applied for a job that needed several years' experience, and unsurprisingly wasn't asked for interview. However, the UK conservation agencies were also recruiting for a series of short contracts for more junior staff and my CV was passed to the recruiters, who got in touch. Looking at my CV, you might think my career was a carefully planned set of steps that came easily. In reality, when I needed to change location, I took over two years to find a job in the right place.

3) As ecologists we never, ever stop learning. One of the most important things is to know when to ask for others' input, whether a specialist, someone with more experience, or a group to brainstorm when faced with something truly new.

4) Find something that inspires you. For me, it isn't so much a question of why I work as an ecologist; in my mind, the question is why everyone else doesn't want to be one! For me, I feel I ended up in my ideal job – it is very interesting, it is important and I feel that I can make a difference. If you have different motivations, that is fine – find something where you can be true to yourself.

5) We are living in a time of intertwined climate and biodiversity emergencies. Ecologists and other environmentalists have a critical role to play. We have a real challenge to give clear, balanced advice to non-experts at a time when it has never been more important. It may be that you end up in a different career, but remember why you care about the environment – whatever job you do, you will be able to take action in your job and in your own time to make a difference, through things like calling for sustainable practices in your workplace.

Personal Profile

I did a degree in pure and applied biology, and then a master's in rural resources and environmental policy. Initially, I worked for government nature conservation bodies and moving into consultancy was purely because I wanted to find a job close to my husband. In the 1990s, there were far fewer jobs for ecologists, and several people working in English Nature (now Natural England), including me, lived over a hundred miles from their spouses and only saw them at weekends. Once I moved into consultancy, I initially worked for a small consultant, and learned a lot there, but I moved to Atkins for the chance to work on a greater variety of projects.

I am an Associate Director of Ecology at Atkins, a large engineering and environmental consultancy. Most of my time is spent advising on the protection of wildlife on major infrastructure projects such as road schemes and railways. I also work on introducing new approaches within ecological and wider environmental consultancy and decision-making, like Biodiversity Net Gain and Natural Capital valuation, where they can benefit nature if introduced alongside existing systems like legal policy and protection.

One of my proudest achievements at Atkins was not a huge engineering scheme, but a small study we did a few years ago for London Wildlife Trust (LWT). We undertook an ecosystem services valuation of Camley Street Natural Park. Using our study as support for applications, LWT secured over £1 million funding for a new visitor centre, due to open in 2020. LWT also used the study as evidence, campaigning to improve protection of locally designated wildlife sites across the whole of London. I used Atkins' innovation fund and inspired colleagues to volunteer time through our volunteering system, so the study was a free donation to LWT.

More information about the project can be found here:

www.atkinsglobal.com/~/media/Files/A/Atkins-Corporate/group/cs/Camley-st-natural-park.pdf

Source: Claire Wansbury, Associate Director of Ecology, Atkins Ltd, UK. © John Wiley & Sons.

Company Profile

Atkins Ltd

Atkins, a member of the SNC-Lavalin Group, is a leading global consultancy with over 80 years of design, engineering and project management expertise. We partner with our clients to deliver complex projects which have a positive, sustainable impact on the world. Our work covers everything from transportation to defence, energy to infrastructure.

With over 18 000 talented employees worldwide, colleagues are surrounded by the skills, expertise and knowledge to help them grow and succeed. Individual's experience and perspectives will play a key role in shaping the projects they work on and Atkins support everyone as they develop their careers.

www.atkinsglobal.com/en-GB

www.linkedin.com/company/atkins

Twitter: @atkinsglobal and @insideAtkins

www.careers.snclavalin.com/ecology-careers

Source: Atkins Ltd. www.careers.snclavalin.com/ecology-careers. © Atkins Ltd.

Personal Profile

Jamie Sneddon, Research Field Assistant at the University of Highlands and Islands, Scotland

Since graduating with a 2:1 BSc in Zoology from the University of Aberdeen in 2015, I've had a very mixed career. My first role involved flying Harris Hawks for a non-lethal pest control company. I had no experience with birds of prey or pest control, but I adapted to the role quickly. Following this, I volunteered with, and later managed, a Scottish wildcat project focused around the trap, neuter and release of feral cats in the Scottish Highlands.

Over the next few years, I spent some time out of the sector, working mainly in education as a social enterprise and pupil support assistant. This break allowed me to build up vital funds in order to gain more experience in the conservation sector.

In 2017, I volunteered as a research field assistant at the University of the Highlands and Islands, working with red squirrels, before securing my first paid role in conservation as a consultant with the Humane Society International in 2018. This work focused on mapping rabies and measuring the success of neutering projects, targeted at stray dogs, around the world. From here, I took on further work as a consultant for Scottish Natural Heritage in 2019 on the Scottish Wildcat Action Project. The role involved analysing camera trap data and identifying individual cats based on their pelage and location.

Presently, I am a research field assistant with the University of the Highlands and Islands again. This role involves radio collaring squirrels to assess the impact of forestry operations on population densities, drey use and dispersal.

My career has only been made possible by taking opportunities as they appear. Many were only available due to making important connections through volunteering. I have often found myself working three jobs at the same time in order to gain as much experience as possible, be able to pay my bills, and to not let any opportunities pass me by. In the five years since my graduation, I have worked tirelessly to cement myself within the sector. It has not been easy, and I am not ashamed to admit that I have considered giving up more than once, but my career now feels like more of a progression than a battle. Stick at it! There is definitely light at the end of the, *very long*, post-graduation tunnel. While I would love to continue working with endangered Scottish species, I have no idea where my career is headed next. However, heading into the unknown is part of what makes the sector so exciting.

◀ **Top Tips**

1) **Education:** To be taken seriously in the industry, it is important to have a degree. It's an opportunity for you to prove that you can work at a higher level, get a 'taste' for academia and learn about the latest software and conservation projects. It is important to understand that a degree is not a free ticket to employment. Once you gain your generalist degree in, for example, 'zoology' you might want to specialise your skills through a master's degree. Again, this *isn't* a free pass. You could choose to start working after your first degree, as I have, but you may need to return to study later in order to progress further in the sector. If university isn't for you, there are alternatives. For example, if you have specialist skills, such as tree climbing, you could find yourself working alongside conservation projects as part of a varied career.

2) **Every ship needs an anchor; find your anchor**: When you leave university, you can find yourself drifting like a ship lost at sea. Seeing a world full of possibilities but possessing no certain direction can be a daunting experience. My advice for you is to find something that *anchors* you. Everyone else may be trying to avoid commitments and are, therefore, able to take opportunities at the drop of a hat, so my advice to you can seem strange. However, once you have a firm anchor point, you have something to build upon. This can mean choosing a species you would like to focus on, a geographical area you want to work, people you would love to work with, whether you would like to work in a lab or in the field, or you could simply be anchored by a relationship. For me, it was a combination of all the above. Commitments aren't a bad thing.

3) **General skills**: Once you decide on a starting point for your career, you need to develop a wide range of skills so that you can move towards it. Build upon the skills you learned in university. This could include Geographic Information Systems (GIS), data analysis programs such as R, Excel, etc. Next, target skills that you don't learn within academia. Learn how to orientate in the field, use GPS devices, identify a range of species, observe those species in the wild, and learn to read field signs. Anyone can attempt to learn skills from a book, but field skills take far more time and experience to learn. They take commitment. So, go out in the rain, the snow and the dark. Become comfortable with the uncomfortable sides of conservation and you will know for sure if you can hack it when it really matters.

4) **Network and ask for help**: This is the one piece of advice that you hear over and over again. It is advice that many find frustrating because they don't even know where to start. However, remember tip number two; be *specific*. Don't just randomly contact everyone that works in conservation. If you want to work with pine martens, then research the field. Contact those people who work with pine martens and the organisations that have projects focused on the species. LinkedIn is an incredible tool for networking, so if you don't have an account, you should. Once you're in contact with experts, ask them if you can meet for a coffee, shadow their work, volunteer on projects or ask if they're ever in need of a contractor. People in the sector are often more than happy to help you but remember that they are busy. Be specific about what you want when you contact someone and understand that they could be slow to respond. Be kind and honest as it really does go a long way. The sector is small, everyone knows each other, and people remember passionate, hard workers. Your name will stand out in a pile of applications if an employer already knows you.

5) **Accept the realities of the job but know your worth**: Even if you follow all the above advice, you will have to give your time for free. This is one of the hardest things to accept when working towards a career in conservation. It's unlike most careers, and, yes, it is not right, it is not fair, but it is the reality of the sector. Maybe one day this will change but for now plenty of people will happily work for free if you refuse to. Work hard, earn your place on projects and stick with it while others give up. We all need to take time out of the sector to take other paid work in order to survive, but this is not giving up, so do not beat yourself up. Pursuing a career where you work long hours, in all weathers, for a low wage, it can feel ridiculous to work so hard and compete with so many for a job where you give more than you get back. Accept that the rewards of the job are a sense of contentment. Being in the field is good for you physically and mentally. You are closer to nature and it is something most people, slumped in offices, strive for. With that said, you also need to know your own worth. If you have learned everything you can from a volunteer post, then move on. Do not let others take advantage of you and stop you from progressing. Why would someone pay you if you're forever happy to work for free?

One more thing...

The conservation sector is uniquely all-consuming. You will work antisocial hours, you won't have the money or time for a social life, and it doesn't take long before your work/life balance is very lopsided. Don't destroy yourself for your career. Make time for your family and friends and try to attend important events. Find time to brush the twigs out of your hair and put on clean clothes every once in a while. Your loved ones will thank you and you'll live longer. Let conservation be your passion, not your prison.
   www.linkedin.com/in/jamie-sneddon1992

Source: Jamie Sneddon, Research Field Assistant at the University of Highlands and Islands, Scotland .
© John Wiley & Sons.

◀ **Top Tips**

Nick Askew, Director, Conservation Careers. UK

So, you want to get a job in conservation? Great! With wildlife in crisis all around the world and numbers of threatened species at an all-time high, the natural world needs your help.

The good news is there are a growing number of jobs available in conservation – it's become a professional industry requiring a diverse and growing range of skill sets. The bad news is it's more competitive than ever before, with 92% of conservationists confirming that it's become tougher to get a job in the last decade alone.

So how do you give yourself the best chance of success? My website Conservation Careers (www.conservation-careers.com) asked 146 professional conservationists from 50 countries to provide their career advice. With a combined experience of 1734 years in the sector, here's what they had to say...

1) **You won't get rich, so you'd better love it.** Yes, there are some well paid jobs in conservation, and no, you won't be on the breadline. However, the reality is most conservationists work long hours, in difficult conditions, and are paid less than many of their friends or family. The reason to work in conservation is because you genuinely want to dedicate yourself to helping wildlife. If you do, you'll have one of the most rewarding careers of all; safe in the knowledge you're helping to make the world a better place.

2) **Get familiar with the jobs that are available.** With the job market expanding all the time, knowing what type of role you'd like to do is one of the hardest steps to take. Start by familiarising yourself with the jobs which are available at sites such as Conservation Careers. Read the job details carefully, and ask yourself the question – how excited would I feel if I did this job day in day out? Once you've decided what you might like to do, make a note of the requirements for the roles. This will give you a good idea of the types of skills and experiences you'll need to acquire over time.

3) **Make things happen for yourself.** Your career will only take off if you create opportunities for yourself and take control. Don't wait for someone else to do it for you. Here's a few ideas which might help: Blog – write about nature, yourself, your experiences,

and let people know you're doing it; say yes to things – attend events, meetings and workshops, go to the pub and talk about conservation. Ask people for introductions and for help – a lot of the time they'll say yes, and; be nice to people – it's a small world, and your reputation is the only thing that counts.

4) **Passion isn't enough, you need experience.** It's not enough to say you love wildlife and are determined to work in conservation – you need your experiences and skills to support this. For many, this means volunteering for organisations you'd like to work for, or in roles which are close to what you'd like to do. One of the best times to get voluntary experience is while studying: join your conservation volunteer group and get involved. Use your holidays to gather relevant and high-quality work experience. Although this is often unpaid work, if you're serious about conservation, you should have the time of your life! You might even be lucky enough to get a paid internship.

5) **Get educated, and don't stop learning.** Conservationists are a clever bunch. When asked what their highest-ranking qualification is, survey respondents stated: doctorate (19%), postgraduate (42%), undergraduate (34%) and School level (6%). The type of qualification depends upon your chosen career path, with PhDs being especially useful for science and research for example. If you're not sure what you'll need, ask people working in your chosen field and read the educational requirements in job descriptions carefully (see number 2 above). It's also important to keep abreast of the latest skills and knowledge by attending training courses, watching TED Talks, and being active in your chosen profession.

6) **Be a professional.** Often called soft or transferable skills, these are invaluable in today's job market. Be a good communicator, manage tasks effectively, accept criticism, be adaptable and reliable, have a good work ethic, get on well with your colleagues and be presentable. They're not looking for Superman, but don't want a Muppet either!

7) **Hone your applications to keep them out of the HR bin.** The reality is most jobs have a lot of applications, so you need to do all you can to keep yours out of the reject pile. First of all, focus on the content; always bespoke your CV/resume/cover letter to each job you apply for, and ensure you highlight the results of your work. Use facts and figures wherever you can and provide clear evidence for each key aspect as outlined in the job description. Having got the content right, check, check and check again that your spelling and grammar are all correct. Many applications end up in the bin for the smallest of mistakes. Ask friends, family and your local careers service to help.

8) **Become great at interviews.** So, you got an interview – congratulations! You must be doing lots of things right. Now is your time to shine. And prepare. You must prepare for interviews and be ready to give confident answers to all the questions that might be coming your way. Use the STAR acronym (Situation, Task, Action, and Response) to help convey your experience. For example, I worked as a conservation volunteer for the Yorkshire Wildlife Trust (Situation). My job was to estimate the Barn Owl population in the Lower Derwent Valley (Task). I did this by visiting 120 farms in the study area and speaking with farmers about their local knowledge of the species (Action). As a result of this work, I found the area to hold the highest density of this species in the United

Kingdom (Response). If you're prone to nerves, practise answering interview questions with your friends or family. Finally, don't forget to sit confidently, breathe calmly and smile. You'll feel better for it.

9) **Be familiar with different cultures and languages.** If you're planning to work internationally, being able to speak different languages and to work within different cultural settings will be an advantage. Languages such as French and Spanish will stand you in good stead.

10) **Stay focused, the first job is the hardest to get.** Once you secure that first job, you're on the ladder and now have control. You decide when you want to move into a new role and can wait until the right opportunity comes your way, safe in the knowledge that you're being paid and building your experience.

11) **Enrol in our free online training.** Learn how to quickly and easily start your career as a professional wildlife conservationist with Conservation Careers. If you're a student, jobseeker or career-switcher you'll learn the golden rule for getting started, the key mistakes to avoid and answer your biggest questions. Find out more at: howtogetaconservationjob.com

Personal Profile

I studied Biology at York University (UK), focusing in on Ecology. During my time as an undergraduate, I was also Chairman of the Conservation Volunteer Group at Uni, so every weekend I'd go out and do practical manual work on local nature reserves, digging ponds, putting up fences, building nesting boxes for birds, whatever it might be. This got me networked into the local conservation community.

My best friend at university was the manager of several national nature reserves, and we lived together through that time. The area was a hotspot for Barn Owls which I came to know very well through my volunteering, so my degree naturally led me into doing a PhD focussed on Barn Owl conservation.

After my PhD, I knew that academia wasn't right for me. And all I knew is I wanted to do something different. I landed a job as a consultant ecologist, doing impact assessments and things like that, then moved into the charity sector, where I stayed for about ten years working for BirdLife International (UK) and BirdLife Pacific (Fiji).

My job was to create conservation projects on the ground across six or seven different Pacific countries, to then craft them into proposals, sell them to donors and secure the money. So once the money came in, I would hand over to conservation staff that would then run and implement those projects. I also worked in communications and marketing.

During my career, I've been asked many times 'how do I get a conservation job', and I felt there should be much better careers support, advice and training for people seeking to work as professional conservationists. This is why I set up Conservation Careers (www.conservation-careers.com) six years ago, and am now helping thousands of people each year to get clear, get ready and get hired in the sector globally. If you'd like our help, please get in touch.

www.conservation-careers.com

Source: Nick Askew, Director, Conservation Careers. UK. © John Wiley & Sons.

Personal Profile

Kristina Lynn, Wildlife Biologist and Content Creator, Canada

I'm a wildlife biologist in British Columbia specialising in applied ecology and coastal ecosystems. My work experience includes researching the impact of linear disturbances on carnivore and ungulate behaviour in Northern Alberta, rehabilitating rescued cougars in the Bolivian Amazon, studying carnivore movement in Southern California's changing human–wildlife interface, and protecting coastal ecosystems in British Columbia.

I create science-based YouTube videos for biologists, environmental scientists, and environmental advocates to understand and interpret the natural world and apply that knowledge to their own lifestyle and understanding of the world.

◀ **Top Tips**

Top tips for aspiring environmental scientists/biologists/ecologists:

1) For jobs outside of academia, job experience is key. Focus on building your resume as early as possible through internships and volunteer work

2) If you're wondering if there's environmental jobs in your area, try a job search exercise: find keywords related to your interests and look on job boards for your city. Note down the mandatory and preferred requirements and try to achieve as many as possible before you graduate and start job searching

3) Build up a LinkedIn profile and reach out to people with similar jobs at companies you're interested in. Don't contact people with the sole intent of getting a job, but send a kind and professional message asking for advice and stories of their experiences. Offer to buy them a coffee and do an informational interview if they're in your city

4) Always work with integrity. This is essential for environmental jobs, particularly in applied ecology. Never compromise your morals – ask questions if something makes you feel uncomfortable

5) Don't discount relationships with your classmates. The people in your environmental classes may one day be hiring managers at local companies.
www.wildbiologist.com
www.youtube.com/channel/UCeoKpQJJGHSYDkSn_bBiKBw
www.instagram.com/wildbiologist/

Source: Kristina Lynn, Wildlife Biologist and Content Creator, Canada. © John Wiley & Sons.

◀ **Top Tips**

Dr Eugenie Regan and Nadine Bowles Newark, UK

Five top tips to developing a career in nature conservation

1) Follow your passion

A career in nature conservation is often a vocation. It is a career we pursue because we want to have impact and make a difference in the world. Luckily, our career success – both personally and professionally – is focused on making a difference. Forget ego and focus on what's best for nature conservation.

Perhaps you are already employed in another field, and are worried about making the jump over to conservation? I found myself in this situation too – after 10 years of success in the retail management sector, I knew I wanted to do something more fulfilling with my career. I had a passion for nature conservation but no qualifications or experience in that area. Going back to the 'bottom of the ladder' was a scary prospect. I decided to go back to full-time education as a mature student and train in environmental management – the best decision I ever made. I haven't looked back since!

These days, my work in conservation takes me all over the world, and my time in retail management seems like a lifetime ago. If you have an enthusiasm for the natural world, then follow it all the way. You'll be amazed at how far you can go when your work is your passion!

2) Get field experience

We live in an incredibly beautiful world. Immerse yourself in that. Get your hands dirty. Experience nature conservation first-hand by getting as much field experience as you can. Volunteer, travel, intern – whatever you can do to build a wide range of field experience. Do you have local conservation organisations that need a helping hand? Volunteer your time with them and learn key skills along the way. Maybe you'd like to travel further afield to gain experience of conservation in exciting places across the globe. This may seem like a pipedream, but it's more common than you think. Don't let finances be a barrier – find a way to make it work in a way that works for you!

Internships are another great way to gain experience and expand your network. Find a scheme through your University or scan the websites of your favourite conservation organisations to keep up with opportunities. If you land the internship, then don't feel intimidated – give it everything you have. Use all your existing skills to excel in the position and absorb all the collective knowledge around you like a sponge!

3) Don't sit still

The most successful career people that we know are busy people. They make their own luck. Don't expect a job to land on your lap. Get out and make yourself busy. But be focussed with how you are spending your time. It's no good spending 5 hours job searching if you only spend 20 minutes creating a CV to apply! Plan your time equally so you can optimise your time on the job search websites, and then get out into the world and connect with people. Attending open lectures or specialised short courses at your local university is a great way to do this.

Keep up to date on the latest articles and journal papers so you can speak with confidence when you get introduced to someone who is an influencer in your field. But make sure you listen too. You may want to tell everybody about how great you are but remember that these people were once in your shoes. Find out all you can about how they made their rise to the top of their field. If you're finding it hard to connect with the right people, then focus on gaining all the field experience that you can. Nature is on your doorstep – so there's no excuse!

4) Use all of your skills and experience

Nature conservation is an incredibly diverse field and requires many skills and expertise. Remember this and draw on your full set of skills. Sure, you need to be confident in your technical knowledge, and demonstrating and maintaining this is very important. But there is a whole heap of other skills which are critical to master for career success. Some people call these 'soft skills' – we prefer Seth Godin's 'real skills' instead.

Are you a good communicator? Do you find it easy to make friends or diffuse difficult situations? These are key skills in people management and can help you positively influence others around you.

Perhaps you are more introverted and thoughtful? Preferring to take a back seat and quietly assess what is going on around you? Your self-control and ability to see the big picture is invaluable in nature conservation projects.

Are you a super-organiser? Have to-do lists? Even lists of to-do lists? Take on roles at work that play to these strengths – organise the social events in your office or create a new filing system. You'll soon be everyone's favourite colleague!

You have a huge array of skills at your fingertips. Draw on these and career success is a breeze.

5) Never stop learning

Think about the most successful person you know in nature conservation. How do you think they managed to rise to the top of their field and gain the status they have now? Do you think they left university with all the necessary skills to achieve the things they have in their career? Absolutely not. Life is one big learning experience, and your career is no different. Read voraciously, attend courses, listen to webinars and lectures. Work on improving yourself every single day and try to learn something from every person you meet. Gather feedback on your performance so you can play to your strengths and confront your weaknesses. Feedback is incredibly important – if you don't know where you are going right and wrong, then you'll never move forward!

Eugenie and Nadine

Dr Eugenie Regan works across four international conservation organisations: UNEP WCMC, IUCN, Conservation International and Birdlife International. Eugenie has wide expertise in nature conservation with a particular focus on connecting biodiversity data providers with end users. She has worked in the private sector, for government and for international NGOs at the local, national, and international level. Nadine Bowles Newark works for UNEP WCMC as a Programme Officer with the Ecosystem Assessment Programme and manages a portfolio of projects including a large biodiversity mainstreaming project in three sub-Saharan African countries. Nadine's work focuses on increasing the understanding of decision-makers regarding biodiversity and its many values in order to mainstream biodiversity into development policy.

Eugenie and Nadine have developed 'We Are All Wonder Women' (www.weareall wonderwomen.com) in their spare time to help women working in nature conservation to be inspired, connected and empowered to create an authentic, fulfilling and happy career. Read their blog for practical and inspirational career advice and tips.

Source: Dr Eugenie Regan and Nadine Bowles Newark, UK. © John Wiley & Sons.

◀ **Top Tips**

Dr Ngaio L Richards, Forensics and Field Specialist, Working Dogs for Conservation, North America

1) It is perfectly all right to feel uncertainty about which avenue(s) to pursue, and to explore various options around that. Some people map their lives out from a very early age. For others, it can take longer or be a lifelong endeavour. The important thing is to strive for balance between not rushing into anything versus dragging your feet for too long.

2) School learning is not for everyone (or always financially feasible), nor does a degree guarantee a secure job, let alone one in your chosen field. Apprenticeships, internships and general applied life experience are hugely valuable. On the one hand, some training and expertise may only be acquired in an academic setting. However, the reality is that certain choices, such as earning a degree, will open doors. That said, there is absolutely no shame in being 'overqualified' for a job if you are genuinely interested in it and feel you would be a good fit. If you want it, go for it.

3) Recognise the significance of work–life boundaries relative to your overall goals. Ask yourself if you fully grasp and are willing to accept certain personal sacrifices in order to do work of a more all-encompassing nature. Be honest as you assess this, and do not reproach yourself if you discover a discomfort with certain sacrifices. This is a reflection of your own preferences and reality, not of your dedication.

4) Every so often, circle back on the following questions:
Am I leading the kind of life I hoped to have?
What are my current realities and obstacles to this?
What do I want to do for work?
What would I love to do, but lack proficiency in?
What am I less fond of doing, but am pretty good at?
Never be hard on yourself for wanting to earn a proper salary, or for having bills to pay and not being able to take that 'dream job'. Conversely, do not equate success with a certain level of earning or accumulation of material possessions as goal posts. You get to delineate your parameters for success.

5) Pushing with all your might to achieve goals is admirable, but it can also be counterproductive and draining. For your sanity, develop a sense of when no amount of pushing will budge the obstacle. Using creative, outside the box thinking, you can often go around the obstacle instead. Recognise that part of the issue may simply be that no trail has yet been carved to what you wish to do. That in itself is an undertaking, but at least it gives you a better idea of what you are up against, and where to redirect your energy. Give yourself some credit for being a trailblazer.

6) Through it all, make sure you allow time to do the things that give you joy, in good company. Enjoy the journey as much as possible; after all it is yours. Laugh abundantly, savour simple pleasures and embrace kindness. Believe in yourself and reach out a hand to support fellow travellers on this same path.

Personal Profile

Ngaio Richards is a proud native of southern Québec, Canada. She has a BSc in Environmental Science (Acadia University, Nova Scotia), an MSc in Natural Resource Science (McGill, Québec) and a PhD in Forensic Science (Anglia Ruskin University, England). Her practical experience spans ecological and contaminant monitoring, wildlife forensics, analytical chemistry, ecotoxicology, collaborative conservation and capacity-building.

Fielding in such far-flung places as the Outer Hebrides, the Aleutian Islands, arctic Alaska and the Cameroonian jungle, she has been lucky enough to share the habitat of winged, scaled and fur-bearing creatures of all shapes and sizes in order to study them. While having an over-riding soft spot for scavengers (esp. vultures and dung beetles) and pollinators, her broad research interests include – in no particular order:

- Developing innovative and non-invasive approaches to monitor environmental contaminants.
- Wildlife forensics and environmental crime: focus on deliberate pesticide poisoning.
- Co-managing, monitoring invasive/non-native and imperilled species for habitat remediation.
- Risk assessment of veterinary drug residues and related practices to wildlife and ecosystems.
- Human health and welfare with regard to: dignity', subsistence, livelihood, human–wildlife conflict.

Along with an extensive list of collaborative publications, she has edited two books and moderates a globally cross-disciplinary online forum. Currently, a conservation detection dog handler and field biologist for the Montana-based nonprofit Working Dogs for Conservation, Ngaio is always happiest in secluded snowy wilderness with a dog by her side.

Source: Dr Ngaio L Richards, Forensics & Field Specialist, Working Dogs for Conservation, North America. © John Wiley & Sons.

Personal Profile

Dr Adrian Cooper – Conservationist and Founder of Felixstowe's Community Nature Reserve, UK

My job requires me to define and develop the operational strategy for Felixstowe's Community Nature Reserve. From a practical perspective, this means that I spend a lot of time listening to members of our local community regarding their aspirations for our work. In doing so, I have to manage their expectations about what is possible. I also cooperate with local media, such as Felixstowe TV and Radio, and local print media, regarding the representation of our work.

The skills needed for this job derive from an understanding of local ecology and biogeography on the Felixstowe peninsula. I also need to have a good understanding of appropriate tools and techniques in strategy development. Finally, I need to be a good communicator – in clear simple English – for a variety of media.

My first degree was in geography, from Loughborough University. My PhD focused on the public engagement with conservation spaces. Since then I have researched, travelled, written and broadcast extensively about the public's participation in conservation in a wide variety of environments. In June 2015, I founded Felixstowe's Community Nature Reserve. Local members have supported my role as Chairman.

My main tip/advice is to learn from local people who have an intimate knowledge of specific habitats. There is no substitute for local knowledge. Equally, you should try and do your own research, so you can question and critically review published work. Never stop learning; and never tire of encouraging local people to take responsibility for nature conservation in their home environment.

Source: Dr Adrian Cooper - Conservationist and Founder of Felixstowe's Community Nature Reserve, UK. © John Wiley & Sons.

Personal Profile

Carol Parenzan, former Middle Susquehanna River Keeper, USA

Carol Parenzan can be found in, on, around or near water, and it was her passion for moving water and watersheds, specifically the Susquehanna River in Pennsylvania (United States), that afforded her the opportunity to become the voice of the river as its Middle Susquehanna RIVERKEEPER®. Licensed by WATERKEEPER® Alliance, a rapidly growing, global environmental advocacy group with more than 250 'Keepers in over 30 countries on 5 continents, she advocates to preserve, protect, and improve the North and West Branches of the Susquehanna River and her tributary watersheds, an area challenged by legacy coal mining, new natural gas exploration and delivery, and aquatic life health issues.

In her youth, Parenzan swam competitively. As a teen and young adult, she paddled hundreds of miles and competed in both flatwater and whitewater events. Her passion for water took her to Penn State University, where she earned an environmental engineering degree with a focus in water. Today, she is an open water long-distance swimmer and paddles an ultralight canoe, which she uses to explore and investigate her watershed. She brings over 30 years of experience in water, wastewater and environmental remediation projects to her position as the Middle Susquehanna RIVERKEEPER®.

Additionally, she is a published children's book author specialising in creative non-fiction for the school and library market. Her 30+ titles include those addressing the environment and she frequently presents programmes in schools and libraries. She is currently raising a Nova Scotia Duck Tolling Retriever, named Susquehanna (Sussey). In addition to being the River's 'Little Keeper' and accompanying Parenzan on field investigations, classroom visits and community events, he will be trained to be a sewage/septic sniffer dog and will work in the field to uncover issues that would ultimately impact the river and the Chesapeake Bay.

◀ **Top Tips**
1) Be involved in community programmes (scouting, paddling clubs, environmental non-profits, etc.) for real-world experience. Encourage youth participation.
2) Seek out college programmes of study that include field experience, research opportunities and internships. Get your feet wet!
3) Volunteer in your local watershed – work can go beyond field work – photography, writing, website, social media, fundraising, grants/funding research, community outreach and education.
4) Become a life-long learner – read, watch environmental documentaries and attend workshops and conferences. Think. Don't take everything at face value. Ask questions. Challenge the source.
5) Follow your heart. Waterkeepers around the world are endlessly passionate about their watershed.
   www.MiddleSusquehannaRiverkeeper.org
   Article for Waterkeeper Alliance – www.waterkeeper.org/magazines/volume-14-issue-2/healing-waters
   Carol Parenzan is the former and founding Middle Susquehanna Riverkeeper, part of the Waterkeeper Alliance community of clean water warriors around the world. Currently, she serves as a sales engineer for Nalco Water, an Ecolab Company, working on clean water challenges with industrial clients.

Source: Carol Parenzan, former Middle Susquehanna River Keeper, USA. © John Wiley & Sons.

6.9   External Resources

UK

British Ecologists
www.facebook.com/groups/847369591966377

UK Ecology Jobs and Courses
www.facebook.com/groups/860839280999125

Conservation Careers Advice – Top Ten Tips on how to get a job in conservation
www.youtube.com/watch?v=bpmyQLg5Mjg
Peter Lynch (2009) 'Wildlife & Conservation Volunteering: The Complete Guide', Bradt Travel Guides
Malcolm L. Hunter, David Lindenmayer and Aram Calhoun (2007) 'Saving the Earth as a Career: Advice on Becoming a Conservation Professional' Wiley Blackwell
HOW TO GET INTO WILDLIFE CONSERVATION. Zoology degree, volunteering, working with animals.
www.youtube.com/watch?v=6cepkBw4iZY

Steph Into Nature
www.stephintonature.com

Mark Carwardine (zoologist and conservationist)

How to get a career in conservation
www.markcarwardine.com/advice/careers-in-conservation.html

James Borrell, Conservation Biologist
www.jamesborrell.com/how-to-get-a-job-in-conservation-and-love-your-work

Europe

Rewilding Europe
www.rewildingeurope.com

European Commission – Species protection
www.ec.europa.eu/environment/nature/conservation/index_en.htm

IUCN – Europe
www.iucn.org/regions/europe

Conservation International Europe
www.conservation.org/places/conservation-international-europe

Society for Conservation Biology – Europe
www.conbio.org/groups/sections/europe

European Outdoor Conservation Association
www.outdoorconservation.eu

Asia

Society for Conservation Biology – Asia
www.conbio.org/groups/sections/asia

Wildlife Conservation Society – Asia
www.wcs.org/our-work/places/asia

Squires, D. (2014), Biodiversity Conservation in Asia. *Asia and the Pacific Policy Studies*, 1: 144-159. doi:10.1002/app5.13

Africa

African Conservation Foundation
www.africanconservation.org

Rhett Butler, Mongabay (2019) Transforming African conservation from old social cause into next-gen growth market
www.news.mongabay.com/2019/09/transforming-african-conservation-from-old-social-cause-into-next-gen-growth-market

Conservation Jobs: Africa
www.facebook.com/groups/1774928439290236

North America

Ecological Society of America
www.esa.org

Conservation International
www.conservation.org
www.conservation.org/about/conservation-international-jobs

Student Conservation Association
www.thesca.org

ESA Community LISTSERVs – ECOLOG-L:
www.esa.org/membership/ecolog

(requires account and login)
A Rocha Conservation Internships
www.arocha.ca/conservation-internship

Paul Greenland and AnnaMarie L. Sheldon (2007) 'Career Opportunities in Conservation and the Environment' Checkmark Books
Government of Canada Job Bank – Conservation Officer in Canada
www.jobbank.gc.ca/marketreport/summary-occupation/3202/ca

Ecojobs
www.ecojobs.com
www.ecojobs.com/environmental-internships.htm

Land Trust Alliance – land conservation jobs
www.landtrustalliance.org/list/land-trust-job-board

National Wildlife Federation
www.nwf.org/Home/About-Us/Careers

Ornithology Exchange
www.ornithologyexchange.org/jobs/board

Biggest Tip for Careers in Wildlife Biology and Ecology – Fancy Scientist
https://www.youtube.com/watch?v=nG8vfFPItks

Conservation Job Board (USA):
www.conservationjobboard.com

The Wildlife Society
https://careers.wildlife.org/jobseeker/search/results

CJB Network
www.cjbnetwork.com/search/category/wildlife-science

How To Become A Wildlife Biologist
https://www.youtube.com/watch?v=kyuP6K0Tu2Y

Ecological Society of America
www.esacareercenter.org

South America

Society for Conservation Biology – Latin America and Caribbean
www.conbio.org/groups/sections/latin-america-caribbean

OECD (2018) Biodiversity Conservation and Sustainable Use in Latin America
www.oecd.org/publications/biodiversity-conservation-and-sustainable-use-in-latin-america-9789264309630-en.htm

The Nature Conservancy
www.nature.org/en-us/about-us/where-we-work/latin-america

Oceania

Australia Conservation Foundation
www.acf.org.au

The Nature Conservancy Australis
www.natureaustralia.org.au

Australian Wildlife Conservancy
www.australianwildlife.org

Australia's Strategy for Nature
www.environment.gov.au/biodiversity/conservation/strategy

Department of Conservation, New Zealand
www.doc.govt.nz

ECO
www.eco.org.nz

NRM Jobs (Australia)
www.nrmjobs.com.au

Global

Conservation Job Board
www.conservationjobboard.com

The Nature Conservancy
www.nature.org/en-us/about-us/careers

Biodiversity Professionals
www.biodiversityprofessionals.org

Careers in Nature Conservation
www.facebook.com/groups/1116238188424571

Conservation Jobs
www.facebook.com/conservationjobs

Conservation Careers
www.conservation-careers.com

Conservation Optimism
www.conservationoptimism.org

Environmentjob
www.environmentjob.co.uk/jobs

Youth at the United Nations
www.un.org/youth

Wildlife Conservation Society
www.wcs.org

BirdLife International
www.birdlife.org

FOR UPDATED AND ADDITIONAL RESOURCES, INCLUDING EXTRA CHAPTERS, GO TO WWW.ENV.CAREERS

References

Conservation Careers (2018) 'Top 17 Conservation Organisations for Conservation Job Seekers' www.conservation-careers.com/conservation-jobs-careers-advice/conservation-organisations (accessed 14 September 2020).

Lucas, J., Gora, E., and Alonso, A. (2017). A view of the global conservation job market and how to succeed in it. *Conserv. Biol.* 31 (6): 1223–1231. https://doi.org/10.1111/cobi.12949.

United Nations Convention on Biological Diversity (2011) Strategic plan for biodiversity 2011-2020. https://www.cbd.int/convention/ (accessed 6 April 2021).

Czech, B. (2008). Prospects for reconciling the conflict between economic growth and biodiversity conservation with technological progress. *Conserv. Biol.* 22 (6): 1389–1398. https://doi.org/10.1111/j.1523-1739.2008.01089.x.

Robinson, J.G. (2012). Common and conflicting interests in the engagements between conservation organizations and corporations. *Conserv. Biol.* 26 (6) https://doi.org/10.1111/j.1523-1739.2012.01914.x.

7

Greening Companies and Corporate Sustainability

7.1 Sector Outline

The practice of companies recording, acting upon and reporting their environmental as well as financial performance is a recent introduction.

Iciar Gallo noted in a 2020 article:

> The increasing global concern of environmental issues is relatively new; that is why Environmental Management Systems appeared just few decades ago.
>
> It was in the 1960s when the environmental movement started to become popular, followed in 1972 by the first United Nations conference on the human environment in Stockholm. But it wasn't until 1992, during the Rio de Janeiro Earth Summit, when the United Nations reached an agreement in order to protect the environment by reducing the negative impacts of business' activities.
>
> The same year, BSI group (British Standards Institution) published the world's first Environmental Management Systems standard – the BS 7750. It provided the basis for the International Organization for Standardization to finally develop ISO 14001 – "Environmental Management Systems – specification and guidance for use" in 1996.
>
> Before its first launch in 1996 of the ISO 14000 series of standards, organizations already voluntarily had been developing their own Environmental Management Systems in order to minimize the negative effects of their processes on the environment. However, they didn't have the necessary tools to compare their environmental practices and impacts until ISO established ISO 14001 as a common framework for comparison.

There are many approaches to, and frameworks for, 'delivering' sustainable development. Marilyn Waite (2013), author of 'Sustainability at Work: Careers that Make a Difference', proposed a systems approach to sustainability, called SURF:

> As a result of analyzing the on-the-ground reality of attempting to transform products and services for sustainable development, the author identified gaps and

Global Environmental Careers: The Worldwide Green Jobs Resource, First Edition. Justin Taberham.
© 2022 John Wiley & Sons Ltd. Published 2022 by John Wiley & Sons Ltd.

incoherencies that rendered a new framework necessary. The SURF framework: supply chain, user, relations, and future, enables a systems-level approach, and subsequently a systems-level impact, for decisions made even on a very microscopic level. Start-ups and large companies, public organizations and private ones alike will benefit from adopting this framework and adapting it to their unique needs. The framework was configured based on an extensive analysis of available definitions, understandings, and methods for implementing sustainable development on a concrete level, as well as through discussions with various industries. . .SURF moves beyond the triple-bottom-line approach to sustainable development to place emphasis on the quadruple bottom line.

What Makes a Company Green?

According to the Green Business network website (2019), 'Green businesses adopt principles, policies and practices that improve the quality of life for their customers, employees, communities, and the planet'.

> Green companies adopt principles and practices that protect people AND the planet. They challenge themselves to bring the goals of social and economic justice, environmental sustainability, as well as community health and development, into all of their activities — from production and supply chain management to employee relations and customer service.
>
> To summarize, a green business goes about business in such a way that is environmentally green, sustainable and most likely with a low carbon footprint. A green business makes itself environmentally green through its buildings, workers, emissions, any packaging it may use and making sure the local community is not affected or polluted.

Corporate Sustainability and Corporate Social Responsibility (CSR)

Deloitte (2021) defined corporate sustainability in the following way:

> Sustainability is responsibility for the impact that the organization exerts on its surroundings, in business, environmental and social terms. Conscious management of the impact translates into lower costs, improved external relations and better managed risks.
>
> Sustainability is skilled positioning of the organization in the economic reality, taking account of the social and economic challenges, environmental opportunities and threats. The awareness that the organization functions within a broader framework, amid complex interrelations with many stakeholder groups, allows it to get ready and make use of the opportunities linked with sustainability.
>
> Sustainability is awareness that each entity is surrounded by stakeholders. Building and cultivating good relations with stakeholders based on engagement and dialogue is crucial, because it not only affects the possibilities to manage risks, but also supports development and gives the organization a competitive edge.

> Sustainability is transformation and development of the organization as well as creation of its long-term value based on innovation as well as intellectual and relation capital.

Andrew Beattie (2019) outlined the three pillars of sustainability and what they mean to businesses:

> Corporate sustainability has become a buzzword in companies big and small. Wal-Mart Stores, Inc., McDonald's Corporation and many of the true corporate giants have named sustainability as a key priority moving forward. Now other corporations are under pressure to show how they plan to commit and deliver their goods and services in a sustainable manner. . .
> Corporate sustainability in investment can fall under the terms ESG for environment, social, and governance or the acronym SRI which stands for socially responsible investment.
> Sustainability is most often defined as meeting the needs of the present without compromising the ability of future generations to meet theirs. It has three main pillars: economic, environmental, and social. These three pillars are informally referred to as people, planet and profits.
> The environmental pillar often gets the most attention. Companies are focusing on reducing their carbon footprints, packaging waste, water usage and their overall effect on the environment. Companies have found that have a beneficial impact on the planet can also have a positive financial impact. . .
> The social pillar ties back into another poorly defined concept: social license. A sustainable business should have the support and approval of its employees, stakeholders and the community it operates in. The approaches to securing and maintaining this support are various, but it comes down to treating employees fairly and being a good neighbor and community member, both locally and globally.
> The economic pillar of sustainability is where most businesses feel they are on firm ground. To be sustainable, a business must be profitable. . .profit at any cost is not at all what the economic pillar is about. Activities that fit under the economic pillar include compliance, proper governance and risk management.

Carbon Neutrality

The environmental pillar of sustainability highlights carbon emissions reductions as an aim for companies. 'Carbon neutrality' is a term often used within businesses.
An EU Press Release (2019) defined carbon neutrality:

> Carbon neutrality means having a balance between emitting carbon and absorbing carbon from the atmosphere in carbon sinks. Removing carbon oxide from the atmosphere and then storing it is known as carbon sequestration. In order to achieve net zero emissions, all worldwide greenhouse gas emissions will have to be counterbalanced by carbon sequestration. . .A carbon sink is any system that absorbs more carbon than it emits. The main natural carbon sinks are soil, forests and oceans. . .

Another way to reduce emissions and to pursue carbon neutrality is to offset emissions made in one sector by reducing them somewhere else. This can be done through investment in renewable energy, energy efficiency or other clean, low-carbon technologies.

7.2   Issues and Trends

This sector is one of rapid growth and developing maturity. Certification of company environmental performance, such as the international standard ISO 14,001, and the realisation by businesses that they have a responsibility to operate within sustainability boundaries has led to growth within the sector in all areas of work.

The organisation Institute of Environmental Management & Assessment (IEMA) reported in 2017:

> Data just released on the uptake of the international Environmental Management Systems standard ISO 14001 shows that the number of certificates issued worldwide has grown 8% over the last year, mirroring the 8% uplift reported last year.
>
> IEMA can report that the data published this week by the International Organization for Standardization (ISO) shows a global year-on-year increase of almost 27,000 certificates between 2015 and 2016. The worldwide total of accredited ISO 14001 certificates reported now stands at 346,190, up from 319,324 in the 2015 data report. 323,023 of those are for the 2004 version of the standard, and 23,167 were awarded against the 2015 version.
>
> ISO 14001 is the world's most used environmental management standard, and globally the second most used business standard, used by companies to manage performance in approximately 200 countries. The data on global uptake of ISO 14001 comes two years into a transition period between the 2004 and 2015 versions, with one year remaining.
>
> The UK has held its position as fourth in the global league table for the fourth consecutive year, with 16,761 certificates. However, the UK's total is down from the 17,824 reported last year – the only nation in the top five to report a year-on-year dip. The UK sits behind China, Italy and Japan and one place ahead of Spain in the rankings.

7.3   Key Organisations and Employers

The organisation GlobeScan organises Sustainability Leaders, the longest running survey of its kind, which, for more than 20 years, has been tracking expert opinions on the evolution of the sustainability agenda and perceptions of leading organisations most responsible for driving it forward. It highlights the following organisations as noted by leaders in the sector in its 2020 survey:

> For the tenth consecutive year, Unilever is most frequently named by experts globally as a corporate sustainability leader, but the list has seen some interesting shifts this year with four new companies recognized for their efforts. After dipping in

2019, mentions of Unilever have rebounded, while Patagonia and IKEA have kept their second and third positions. Thirteen companies are mentioned by at least three percent of experts, with four new companies reaching this threshold this year – Microsoft, Ørsted, L'Oréal and Tata.

Unilever, Patagonia and IKEA remain among the top five most frequently mentioned sustainability leaders on all continents, but experts also recognize local companies headquartered in their regions. Natura &Co tops the ranking in Latin America, and Interface, Microsoft and Tesla are among the most mentioned in North America. Similarly, Danone and Ørsted are well ranked in Europe, while Tata is gaining traction in the Asia-Pacific region.

Sustainability professionals point to WWF as the leading NGO advancing global sustainable development, with a spike in recognition since last year. While perceptions of Greenpeace remain steady, the World Resources Institute strongly reinforces its third position compared to 2019.

Newsweek has a regular annual Green Ranking listing of top 'green companies':
Its 2017 ranking included:

---

1) L'Oréal SA

2) Centrica PLC

3) Enbridge Inc.

4) Siemens AG

5) Cisco Systems Inc.

6) Henkel AG & Co KgaA

7) Accenture PLC

8) BT Group PLC

9) Adidas AG

10) Koninklijke Philips NV

Newsweek green rankings are world recognised for the assessment of corporate environmental performance. These companies have been ranked on corporate sustainability and environmental impact.

In their '2019 Global 100' of the Most Sustainable Corporations in the World (a ranking of companies which generate more than $1 billion in annual revenue), Corporate Knights (2019), reported by CSO Magazine (2020), have a top ten as:

---

1) **Chr. Hansen Holding A/S**
This bioscience company has a key focus on the production of cultures, enzymes and natural colourings for the food industry. Chr. Hansen say that its 'purpose is to deliver natural innovative solutions that address global challenges by advancing food, health and productivity'.

2) **Kering SA**

   Kering, the Paris-based luxury goods company, is owner of brands of the likes of Gucci, Yves Saint Laurent and Balenciaga. One factor propelling them up the list was the high percentage of women on its board; 64%, the highest of all the companies examined by Corporate Knights.

3) **Neste Corporation**

   Neste are a Finnish oil refining company. While the presence of an oil refining company on a sustainability list may at first seem a surprise, Neste says: 'We have always thought in a different way. We want to challenge the oil refining industry by offering increasingly clean fuel solutions and applications based on renewable raw materials'.

4) **Ørsted**

   Ørsted is Denmark's largest energy company. Its CEO, Henrik Poulsen, has said 'to slow down climate change, we must transform society with CO2 neutral solutions'. Accordingly, the firm has made a commitment to fully phase out coal by 2023 and have established a focus on green energy such as offshore wind to replace that energy source.

5) **GlaxoSmithKline PLC**

   In September 2018, the London-based pharmaceutical company announced a set of commitments around responsibility, aligning itself with UN Sustainable Development Goal 3 of promoting Good Health and Wellbeing.

6) **Prologis Inc.**

   Prologis is a multinational real estate investment trust headquartered in San Francisco. The company says that it 'strives to be an exemplary corporate citizen, to minimise our environmental impacts and to maximize beneficial outcomes for our stakeholders'.

7) **Umicore**

   Umicore is a Belgian materials company. Having originated as a mining company, Umicore now focuses on clean technologies. Umicore says its 'overriding goal of sustainable value creation is based on an ambition to develop, produce and recycle materials in a way that fulfils its mission: materials for a better life'.

8) **Banco do Brasil S.A.**

   Banco do Brasil describes itself as the largest financial institution in Latin America. Founded in 1808 in Rio de Janeiro and mainly focused on the Brazilian market, the company facilitates access to the country for foreign companies looking to market there.

9) **Shinhan Financial Group Co.**

   South Korean Shinhan Financial Group oversees both Shinhan Bank and the smaller Jeju Bank. Headquartered in Seoul, the organisation is descended from Korea's oldest bank. The company emphasises the role of 'Compassionate Finance' in the way they do business, with CEO Cho Yong-byoung saying, 'going forward, all Shinhan members will practice "compassionate finance for the future" to realise the Group's mission of creating a virtuous cycle of shared prosperity'.

10) **Taiwan Semiconductor**

Taiwan Semiconductor are the world's largest dedicated semiconductor foundry. The firm pioneered the pure play business model for semiconductor foundries, saying: 'By choosing not to design, manufacture or market any semiconductor products under its own name, the Company ensures that it never competes directly with its customers'.

In 2019, the Climate Coaching Alliance was formed to bring together coaches and coaching organisations to share resources, develop strategies and develop dialogue.

7.4 Careers

There are many areas of work which support companies in their programmes for reducing their impacts and becoming more sustainable in their operations. These include environmental auditing, consultancy and advisory services, scientists and engineers assisting in measuring impacts, analysts, programme and operations managers.

Environmental auditing is a relatively new and fast-growing career area and the work can be varied. Auditors may focus on just one area, for example waste minimisation, or conduct more comprehensive general environmental reviews. The work can also involve looking at the management structure of organisations to make sure that the commitment, resources, training and allocation of responsibilities exist to put an environmental management system in place.

It must be noted that you can be a 'green worker' within a 'less than green' company and make as much, or a bigger, impact than someone who works for one of the top green companies.

7.5 Job Titles in the Sector

Sustainability Manager, Sustainability Director, Environmental Auditor, Sustainability Engineer, Sustainability Specialist, Asset Manager, Project Manager, Strategy Specialist, Sustainability Consultant, Carbon Consultant, Energy and Renewables Consultant and Energy Analyst.

7.6 Educational Requirements

Emily Chan (n.d.), a sustainability strategist based in San Francisco, outlined her views on key educational requirements in her article for The Muse:

As a corporate sustainability consultant at GreenOrder, the most common question I received was, "What type of graduate degree do I need to be a competitive candidate in this field?"

The first thing I'd say? "Slow down." Setting yourself up for a successful career in sustainability is about more than the graduate degree you get. In fact, I think that taking time to try some things out first is critical in helping you choose the right grad program — and ultimately landing a job you love. . .

Step 1: Don't Go to Grad School

That's right: The first step in a successful sustainable career is to not go to grad school (at least not right away).

Why? Because before you can really decide what you want to devote your time (and money) to, you need to get your hands dirty and learn what you actually like. . .look for an opportunity that gives you a chance to experience two or three different functional roles.

Step 2: Learn Which Job Functions You Like

While you're working, it's important to pay close attention to what you like and don't like about different jobs and tasks.

Step 3: Choose Which Grad Program is Right

While grad school certainly isn't required for a successful career in sustainability, it's an important path to consider. . .Most professionals working in the field either have a master's degree or have already had about 10 years of highly regarded work experience. . .

Ready to launch your green career? Having a few years of work experience under your belt, knowing which functional areas you excel at naturally, and selecting a complementary graduate degree will not only set you up for a job in sustainability, it will set you up for an impactful career you're passionate about.

Study.com highlighted the differing needs for varied roles in the sector in its 2019 article 'Environmental Sustainability Career Information and Options':

> People who are interested in pursuing a career in the field of environmental sustainability usually need to have at least a bachelor's degree and some related work experience. Applicable degree fields include environmental sustainability, environmental management, management or engineering, among others. . .Jobs for environmental sustainability management positions require experience in the field as well an understanding of public administration and policy. In addition, you should enjoy working both in an office and outside.

There are many degrees and higher-level qualifications which are relevant to a career in the sector, which include Environmental Sustainability, Business Management and Sustainability and Sustainable Finance.

7.7 Personal Attributes and Skill Sets

Personal attributes and skills which are helpful in the sector include:

- Communication and interpersonal skills
- Data and article analysis

- Project management
- Report writing, including technical and corporate
- Client management
- Researching
- Writing for different audiences
- Knowledge of legislation, regulations and working practices
- Media, PR and communications
- Scientific knowledge within the sector

Training

Center for Sustainability and Excellence
www.cse-net.org

The Sustainability Academy
www.sustainability-academy.org/about-us

edX
www.edx.org/course/introduction-to-corporate-sustainability-social-in

Coursera listing of sustainability courses
www.coursera.org/courses?query=sustainability

CSR Works
www.csrworks.com/training

Deloitte
www2.deloitte.com/au/en/pages/risk/solutions/sustainability-corporate-social-responsibility-training.html

7.8   Career Paths and Case Studies

Personal Profile

Sharmila Singh, Founder and CEO of New Lens Consulting, USA

I recall my first ecology class in undergrad; it forever changed the way I viewed the natural world and the interconnectivity among ecosystems. While I began my career in business, it was the memory of this class that influenced a career pivot in 2004. I joined the Presidio Graduate School where I received my MBA in Sustainable Management. During my time there, I took on internships and engaged in experiential learning projects to gain my experience in sustainable operations, global supply chain, effective storytelling and communications. I worked to support the first Green Chamber of Commerce in San Francisco, designed the first 'green' fashion show for SF Fashion week, created a my first SROI (Social Return on Investment) report for a fair-trade chocolate company and increased engagement for a Green Training Program developed by Goodwill Industries.

After I graduated, I started consulting and began helping Fortune 500 companies implement social and environmental initiatives and establish sustainability action plans. This involved identifying core sustainability initiatives, setting benchmark metrics for water, waste and energy efficiencies and engaging teams to implement best practices for social and environmental responsibility. Companies I have worked with include Clif Bar, Constellation Brands, Sephora, Amy's Kitchen, Lagunitas Brewing, Jackson Family Wines and Marriott Bonvoy. I found my true calling helping organisations commit to growing their business in accordance with the Triple Bottom Line of People, Planet and Profit.

My own career journey has charged a desire to help others discover the career paths that bring them deep joy and fulfillment. I founded New Lens Coaching to support professionals who are striving to work for mission aligned companies and those who are ready to advance in the field. I use frameworks and techniques to support clients in their career search, networking, building an online presence, designing effective resumes and mastering interviewing skills.

Website: www.newlensconsulting.com

LinkedIn: www.linkedin.com/in/sharmila-singh-mba/

Email: sharmilas@newlensconsulting.com

Five top tips for people seeking a career in the sector.

Tip One: Identify What You Do Well. Many sustainability professionals make a career pivot into the sustainability field from other industries with backgrounds in a variety of expertise such as program management, communications, operations and accounting. These are transferable skills that will serve you well as you build your career in sustainability and can be your entry point into the field. The path is not always linear. You can seek out roles that make a direct impact or effectively land a role in your area of competency and *then* engage with the sustainability efforts within the company.

Climate change is the biggest, most important and most complex adventure humans will face. Take what you do best and apply it to this cause. I guarantee you'll find fulfillment, fascination and a differentiation that puts you at the forefront of an emerging field.

Whether you're a lawyer, educator, engineer, artist…become a climate lawyer, climate educator, climate engineer, climate artist. Explore what that intersection means for you. Start by thinking, 'how can all facets of a company and or organization think more sustainability?'

Tip Two: Find Your Niche.

In the job market we are seeing the need for professionals with an expertise in specific areas such as green building, zero waste, circular design, global supply chain etc. Finding the path that you wish to take can be daunting, so start by asking yourself, 'What sustainability issues am I most passionate about and what kind of impacts do I want to make?' Approaching your efforts in terms of impacts can help you shape your career. You can refer to the UN SDG's to identify the core areas of impact you would like to engage in or follow publications such as GreenBiz and Sustainable Brands to get an idea of what topics pique your interest. The 80,000 Hours job board site is an excellent resource for those looking for a career as it lists global opportunities by the problem areas you wish to address in your career.

Tip Three: What Am I Qualified for?

While this is not always the case, some sustainability roles require specific certifications, schooling and years of experience. One great way to assess where you stand is to review the role descriptions for the types of roles that you are striving for, taking note of the skills that you currently possess and those which you may need to acquire by experience, education or additional credentials. While certifications and education can be important to uplevel your career, do not ignore the need to gain experience in implementation. Teams look for candidates with both the education and the 'on the ground' experience working in the field. You can attain these skills via internships, fellowships and by incorporating sustainability in your current role.

Tip Four: Build Your Network Online. The field of sustainability is continuously innovating and evolving. Stay in the know of what is happening by 'following' companies and organizations on LinkedIn so you receive updates regarding the latest news and upcoming opportunities as jobs are often posted via LinkedIn Company Pages. Follow LinkedIn groups that are focused on sustainability and follow hashtags that speak to your interests such as #impactinvesting, #sustainablefashion and #sustainabledevelopment to be alerted of content posted in these areas. Utilize LinkedIn to build your own online brand by re-posting the articles of relevance and sharing your insights about the content, showcasing your own thought leadership.

Tip Five: Connect. You can learn so much by hearing from like-minded individuals who are working in the field. Use LinkedIn to connect or follow those who inspire you. Take the time to send a short note as to why you are connecting and what your interests are. Go one step further and ask for a 20-minute conversation to learn more about the initiatives they are currently working on, their future goals and any insights that they have relating to the field. Get specific regarding your 'ask' and what you find most interesting about the work they are doing. These conversations should be mutually beneficial, so ask how you may support them or take the initiative by sharing content that you find interesting and other resources that may be beneficial.

Source: Sharmila Singh, Founder and CEO of New Lens Consulting, USA. © John Wiley & Sons.

Personal Profile

Trish Kenlon, Founder, Sustainable Career Pathways, USA

After working for several years in operations and consulting for large corporations such as Bank of America, IBM, and CGI, I wanted to pivot to a career that would allow me to use my business skills and experience to help solve the climate crisis. After doing a lot of research, I realised that an MBA at New York University's Stern School of Business was the best way to make that transition.

While at NYU, I pursued every learning opportunity that I could – I led our Net Impact chapter, I volunteered for our school sustainability office and I used class projects to deepen my understanding of specific issues and tools such as carbon accounting, waste to energy and B Corp certification. I also networked extensively – I attended conferences, joined professional groups and had coffee with sustainability practitioners all over the east coast.

My hard work paid off, and I was selected to join the second-ever cohort of the Environmental Defense Fund's Climate Corps fellowship at TXU Energy. The EDF fellowship gave me the experience I needed to become ANN INC.'s first Manager of Energy and Sustainability.

A few years ago, I realized that people had been referring their friends to me for guidance on how to get started working in sustainability for over a decade and that it was time to start documenting my knowledge. It quickly became clear that there was so much material I wanted to share with people that I had to create a website, which became SustainableCareerPathways.com. Once it went live, the response was so overwhelmingly positive that I realised that I'd created my own dream job – helping others to find their place in this vast and often confusing world of sustainability careers.

Trish Kenlon is the founder of Sustainable Career Pathways (SCP), a popular website and sustainability career coaching company. Her practice focuses on helping graduate students and mid-career professionals with transitioning into roles in the sustainability and impact space. She is a frequent contributor of career expertise to the EDF Climate Corps community and has presented on techniques for an effective impact job search at Carnegie Mellon University, the University of Miami, Johns Hopkins University and the Association of Energy Services Professionals. Prior to SCP, Trish worked in sustainability roles for Ann Taylor, the Environmental Defense Fund and TXU Energy. She's also worked in project management and consulting roles for IBM, CGI and Bank of America. She is an alumna of Carnegie Mellon University, the NYU Stern School of Business and the EDF Climate Corps. Website: www.sustainablecareerpathways.com

◀ Top Tips

1) Begin by narrowing the focus of your job search. Reading books such as *Global Environmental Careers*, or Katie Kross's *Profession and Purpose*, and visiting sites like www.sustainablecareerpathways.com can give you a better understanding of the opportunities in the field. Once you've targeted the work you want to do, find some examples of job descriptions and use them to help build your story about what you're looking for and the value you can add. Having that written down on paper will help you prioritise what conferences to attend, what networking groups to join, what job boards to frequent and who to reach out to.

2) While it's important to figure out what direction you are headed in to help you develop your story and guide your search, try to also stay open to opportunities that you may come across in your networking even if they don't match your target criteria 100%. The reality is that sometimes the right job for you is the one that falls in your lap and meets enough of your original criteria to make you happy.

3) While some people have actually had success with job boards (the SCP Job Resources page has over 30 great ones!), the truth is networking is probably going to be your best bet. The 80/20 rule really rings true here; you should spend about 20% of your time applying to jobs online and 80% of your time networking at events, asking for informational interviews and reaching out to people via LinkedIn.

4) Before getting out there and making a wonderful impression on everyone you meet, make sure you've got your marketing materials ready. Invest some time in making your LinkedIn profile shine, tighten up your resume so that it highlights the skills and expertise you want to showcase, write a solid cover letter that you can tailor as needed and perfect your elevator pitch.

5) If you believe a career in sustainability is truly what you want but you have absolutely no experience, go out and get some, just know that it probably won't be paid (at least at first). Ideally you would go to graduate school or find an internship or fellowship that would pay you while you learn, but if that's not possible, volunteering is a fantastic way to get some experience with sustainability issues while building your network. Look for volunteering opportunities with non-profits, your company's sustainability team, your school's sustainability office, local professional groups or local advocacy groups.

6) I've had clients that have met virtually every criteria in a job posting except for one, and they still wouldn't apply for that job. It's hard because you don't want to waste your time applying for a job you won't get...but is that necessarily the outcome? If you find a job posting that looks like something you know you'd be great at but you don't fit a good chunk of the requirements, apply anyway! It's usually pretty hard to find postings that fit what you're looking for, so if you find one that even kind of sounds good, it's worth trying. The worst thing they can do is not hire you, and that's going to happen anyway if you don't apply. You never know when a company's actual key requirement is something like fit with the culture, willingness to learn or something else you can't put on a resume.

7) The majority of sustainability jobs that are out there are hiring to fill an immediate need so if you're a student, that means you very well could be graduating without a job. This is a particularly hard pill to swallow when all your banking and consulting friends get jobs in January and spend their last semesters on the beach. That's not to say you should wait until spring to start your job search – figuring out what you want, refining your positioning, networking and building relationships take a lot of time, and your goal is to be the first person people think of when a new job opens up at their organisation. That doesn't happen overnight, so get started!
   Website: www.sustainablecareerpathways.com

Source: Trish Kenlon, Founder, Sustainable Career Pathways, USA. © John Wiley & Sons.

Top Tips

Katie Kross, Author, 'Profession and Purpose: A Resource Guide for MBA Careers in Sustainability', USA

1) There are lots of ways to use your skills to have a positive impact on the world. To focus your search, ask yourself five questions: What is the issue/impact area you want to work on? What sector do you want to work in (big company, start-up, non-profit, government)? What industry (e.g. apparel, tech, banking, food and agriculture)? What job function do you want to hold (marketing, finance, communications, energy management,

sustainability strategy)? And, finally, what is the geographic focus of your search? There are meaningful sustainability careers in all sectors, industries, functions and locations.

2) More than half of sustainability professionals have their jobs as a result of a networking conversation (not a public job posting). Focus the majority of your search time on building relationships with industry professionals at the organisations you're most interested in. Make a list of the top 30 organisations you're most interested in working for, then systematically work on finding connections and requesting conversations at those organisations.

3) If you want to work in the private sector, be prepared to make "the business case" for sustainability in an interview. Why is sustainability good for a company's bottom line, not just the earth? Can you put it in the context of that specific company/industry/product line?

4) This field is evolving rapidly. Many of the sustainability jobs that exist today did not exist even five years ago – and the same will be true for the next five years. Stay current on industry news as well as sustainability and climate news. Think ahead. Look for opportunities to invent a role for yourself.

5) There are so many ways to make a positive difference in an organisation. Sometimes the most important changes come from professionals who do not have sustainability in their job titles but are in a position to directly transform the business. If you work in operations, can you propose a new energy efficiency strategy? If you work in marketing, can you suggest new ideas for plastic-free packaging? Can you volunteer to start a new employee engagement program? A bike to work day? Ask for a socially responsible mutual fund choice in your company's 401k option set? Look for opportunities to lead from wherever you are.

Twitter: @Katie_Kross

Source: Katie Kross, Author, 'Profession and Purpose: A Resource Guide for MBA Careers in Sustainability', USA. © John Wiley & Sons.

Personal Profile

Chhaya Bhanti, Vertiver, India

Chhaya Bhanti is the Founder and Creative Director of Vertiver, a sustainability and design consultancy and Co-Founder of Iora Ecological Solutions, and works with policy makers, researchers, companies and organisations to manage knowledge, build communication tools and offer environmental research and advisory on various issues across waste, water, climate change, forests, biodiversity etc. A key skill required in sustainability is to help various stakeholders comprehend the complexity of a given environmental/social issue and help them take action accordingly. In order to create the tools that enable this understanding, you need to have systems thinking skills so that you can connect the social, economic, technological, political and environmental aspects of an issue into an easy-to-understand model. If you want to play this role, you need to develop an understanding of how visual communication overlaps with data science, how research findings can be disseminated in language that is not weighed down by obtuse jargon and how to develop maps of processes and partners that may be part of a solution to a sustainability issue.

If you have a degree in social sciences, arts, engineering, business administration or many other non-environmental fields, you can still be very valuable for an environmental company if your heart truly cares about conservation. We always like to hire people based on their personal passion about sustainability. It comes across in the issues they care about, previous volunteer or other related work they have done, how they might work to solve an issue and how aware they are of current trends in that area of work. While it's smart to be a specialist in a particular area of sustainability, since all projects require a very collaborative approach with many experts working together, it is also good to acquire the skills of being a generalist. It's good to know for instance, how climate change affects forests and vice versa, how groundwater is affected by waste or how water scarcity is connected to livelihoods etc.

Once you begin to collect knowledge of an area of sustainability you really care about, such as forests, biodiversity, climate change, waste management, water etc., you will yourself begin to connect the dots on the role you may want to play in that area. Ask yourself whether you would ideally like to manage a sustainability project on the field or would you rather write about it, would you like to work one-on-one with communities on the ground or would you rather conduct primary/secondary research and publish those findings. . .there are many ways you can contribute to the sustainability sector once you identify your entry point of interest.

Sustainability is a very large and complex field, and while your heart may be sincerely focused on solving the big issues of the planet, the first step you need to take is to map out a small sphere that interests you deeply and overlay that with your skills. See how your current skills match the needs and work profile of organisations that excel in that area of sustainability and then assess what you need to add on. Always continue building your writing portfolio as well. Write about areas that interest you. Employers love good writing. It helps them manage projects more efficiently with clients and also communicate the many aspects of the project to a larger audience to create maximum impact. The world of sustainability needs design thinkers, strategists, technologists, writers, photographers, storytellers, data and other scientists to work together collaboratively. Applying your skills to restoring our social and environmental ecosystems is the most fulfilling job you'll ever do.

Company Profile – Vertiver

Vertiver is a multidisciplinary communication, behavior change and knowledge advisory platform. They work with policy makers, development agencies and organisations to develop and communicate action-oriented solutions on various aspects of social and environmental sustainability. They excel in translating research into easy-to-comprehend awareness and knowledge tools, engaging with stakeholders to develop and implement projects, conducting ethnographic research and developing and implementing behavior change programs through multi-sectoral partnerships.

Website: www.vertiver.com
Twitter: @VertiverAgency
LinkedIn: www.linkedin.com/company/vertiver

Source: ChhayaBhanti, Vertiver, India. © John Wiley & Sons.

◀ **Top Tips**

Marilyn Waite, William and Flora Hewlett Foundation

Personal Profile
Marilyn Waite leads the climate and clean energy finance portfolio at the Hewlett Foundation, covering the markets of China, Europe and the United States. An avid writer and communicator, she is author of *Sustainability at Work: Careers That Make a Difference* and The Innovators column on GreenBiz. Marilyn also serves on various clean energy and investment boards, including the Kachuwa Impact Fund.

Marilyn previously led the energy practice at the venture capital firm Village Capital, modelled and forecasted energy solutions to climate change as a senior research fellow at Project Drawdown and served in a number of roles in nuclear and renewable energy at Orano (formerly AREVA). She taught sustainable business at the University of International Business and Economics (UIBE) in Beijing and worked in a number of capacities throughout China.

She holds a master's degree with distinction in Engineering for Sustainable Development from the University of Cambridge and a Bachelor of Science Degree in Civil and Environmental Engineering, magna cum laude, from Princeton University.

Website: www.marilynwaite.com

Source: Marilyn Waite, William and Flora Hewlett Foundation. © John Wiley & Sons.

Personal Profile

Tiernan Humphrys

Tiernan is a seasoned sustainability professional with over 17 years' experience in both the public and private sectors in Ireland, England and Australia.

He studied environmental management and town planning for nine years in Ireland, Scotland, Canada and England, winning awards at both the University of Aberdeen and University of the West of England.

He started his career as an environmental consultant with Arup working on major infrastructure projects. In 2002 he moved to the public sector and has forged a career in sustainability in government. He worked in the cross government Sustainable Development Unit, during which time he represented the British Government at a UN conference on sustainable public procurement.

In 2004 he migrated to Australia continuing his speciality of sustainability in government. He has worked for the Victorian State Department of Sustainability and Environment, VicRoads and the Office of the Commissioner for Environmental Sustainability.

Since 2008 he has been Manager Environmental Sustainability at the Department of Health and Human Services where he is responsible for providing sustainability leadership to the state's 87 public health services, which employ some 80,000 staff, covers over 3 million square metres and emits over 750,000 tonnes of carbon. During this time total water use has reduced by close to 1 billion litres, energy intensity per square metre and occupied bed-day has reduced, over 1.5 megawatts of solar power has been installed, over 90% of health services have an environmental management plan and over 80% are publicly reporting on their environmental performance.

In 2015 his team received the inaugural Institute of Public Administration Australia (Victoria) Leadership in Sustainability Award. For the last three years his contribution to sustainability has been recognised through judging the Victorian Premier's Sustainability Award.

In 2016 he was awarded a scholarship from Yale University to attend their inaugural Sustainability Leadership Forum.

'I got my first job at Arup following a work placement which I secured by phoning the Managing Director of the company until he took my call. It started out as a placement for a few weeks; I ended staying for the entire summer holidays and was then offered a full-time role after university. During the placement I tried as many different things as I could – geotech, aerial map reading, editing and even spent time in their print room. So, create opportunities, be prepared to try new things and make the most of every opportunity – they don't come around very often.

It is also important to understand why you want to work in the environmental sector.

If you want to change the world join an NGO (but the pay isn't great), if you want to make money join the private sector (but be prepared to defend things you may not agree with) and if you want to work in policy join the public sector (but remember you work for the Government of the day). The public sector is I think underrated – there are huge career opportunities given the breadth of work, the work can be really interesting and job security is always a bonus.

The skills I have valued most are writing, interpreting technical issues and being able to condense key issues succinctly (Ministers generally haven't time to read more than a page). Knowing a bit about everything is important; being what I call specialist generalist. My current role can cover issues such as organics, cogeneration, energy efficiency, solar, PVC recycling, water recycling and climate change in a day so being able to easily grasp and switch between issues is a must'.

Source: Tiernan Humphrys. © John Wiley & Sons.

7.9 External Resources

UK

Sustainability First
www.sustainabilityfirst.org.uk

Sustainability Exchange
www.sustainabilityexchange.ac.uk/home

Lewis Davey
www.lewisdavey.com

Prospects Job Profile – Sustainability Consultant
www.prospects.ac.uk/job-profiles/sustainability-consultant

Sustainababble podcast
www.sustainababble.fish

Cheeky Panda
www.thecheekypanda.com/UK

Europe

European Commission – Sustainable development
www.ec.europa.eu/environment/sustainable-development/index_en.htm

Sustain Europe
www.sustaineurope.com

CSR Europe
www.csreurope.org

Research Report by Board Agenda & Mazars in association with INSEAD Corporate
Governance Centre 'Leadership in Corporate Sustainability – European Report 2018'
www.insead.edu/sites/default/files/assets/dept/centres/icgc/docs/leadership-in-
corporate-sustainability-european-report-2018.pdf

Asia

BSR 'Sustainable Business in Asia: Five Trends That Will Impact the Decisive Decade'
www.bsr.org/en/our-insights/blog-view/sustainable-business-in-asia-five-trends-that-will-
impact-decisive-decade

World Business Council for Sustainable Development (WBCSD), in partnership with
the Climate
Disclosure Standards Board (CDSB) and Ecodesk – The Reporting Exchange 'Corporate
and sustainability reporting in Singapore and Southeast Asia'
www.docs.wbcsd.org/2018/10/Corporate_and_sustainability_reporting_in_Singapore_
and_Southeast_Asia.pdf

Merriden Varrall, Asia Society (2020) 'ASEAN's Way to Sustainable Development'
www.asiasociety.org/australia/aseans-way-sustainable-development

Africa

David Craig, World Economic Forum Agenda (2019) 'It's in the numbers – how sustainability will support Africa's long-term growth and development'
www.weforum.org/agenda/2019/09/why-sustainability-is-critical-for-the-long-term-growth-
and-development-of-africa

Africa Sustainability Matters
www.africasustainabilitymatters.com

United Nations (2019) 'Corporate Sustainability Reporting in Least Developed Countries:
Challenges and Opportunities for Action'
www.unohrlls.org/custom-content/uploads/2019/03/Occasional-Paper-6.pdf

United Nations Global Compact – Africa
www.unglobalcompact.org/engage-locally/africa

North America

Ed's Clean Energy & Sustainability Jobs List
www.edsjobslist.com

The Environmental Career Center
www.environmentalcareer.com

B Work connects purpose driven jobseekers with meaningful work at companies that are using business as a force for good.
www.bwork.com

Sustainable Career Pathways, network of organisations
www.sustainablecareerpathways.com/networks

Sustainablebusiness.com
www.sustainablebusiness.com

Climatebase
www.climate.careers

GreenBiz Group
www.greenbiz.com

GreenBiz Jobs Board
www.jobs.greenbiz.com

Net Impact
www.netimpact.org

Net Impact Careers
www.netimpact.org/discover-your-career
www.netimpact.org/jobs

80,000 Hours aims to solve the most pressing skill bottlenecks in the world's most pressing problems
www.80000hours.org

The Sustain O'bility One Stop
www.sustainobility.biz

(subscriptions to gain help with career development)
How I got my sustainability job
www.youtube.com/watch?v=JSOLbfwYCTo

Career paths in sustainability
www.youtube.com/watch?v=u1uQk7UVuvA

Weinreb Group Sustainability and ESG Recruiting
www.weinrebgroup.com

Corporate Knights, one of the world's largest circulation magazines focused on the intersection of business and sustainability
www.corporateknights.com

Auden Schendler (2010) 'Getting Green Done: Hard Truths from the Front Lines of the Sustainability Revolution'

Amory Lovins and Rocky Mountain Institute (2011) 'Reinventing Fire: Bold Business Solutions for the New Energy Era'

Rocky Mountain Institute
www.rmi.org

Rocky Mountain Institute Careers
www.rmi.org/about/careers

International Society of Sustainability Professionals (ISSP)
www.sustainabilityprofessionals.org
www.sustainabilityprofessionals.careerwebsite.com

Green America
www.greenamerica.org

South America

World Business Council for Sustainable Development (WBCSD) (2020) 'Latin America'
www.wbcsd.org/Overview/Global-Network/Regions/Latin-America

IndexAmericas (Inter-American Development Bank)
www.indexamericas.iadb.org

Grantham Institute (2019) 'How can Latin American countries shape a more sustainable future?'
www.granthaminstitute.com/2019/09/02/how-can-latin-america-countries-shape-a-more-sustainable-future

Oceania

Australian Council of Superannuation Investors limited (ACSI) (2018) 'Corporate Sustainability Reporting in Australia: 2018'
www.acsi.org.au/research-reports/corporate-sustainability-reporting-in-australia-2018

World Business Council for Sustainable Development (WBCSD) (2018) 'Sustainability reporting in Australia: jumping into the mainstream'
www.wbcsd.org/Programs/Redefining-Value/External-Disclosure/The-Reporting-Exchange/News/Sustainability-reporting-in-Australia-jumping-into-the-mainstream

Sustainable Business Network
www.sustainable.org.nz

Global

Green Jobs
www.greenjobs.co.uk/browse-jobs/sustainability-jobs

Green Recruitment Company
www.greenrecruitmentcompany.com

Josh's Water Jobs

www.joshswaterjobs.com/jobs

Earth Hackers – monthly newsletter focusing on three key companies

www.earthhackers.substack.com/archive

Climate Coaching Alliance

www.climatecoachingalliance.org

CSO Magazine

www.csomagazine.com

Ibex Earth – 'Student' section includes information on internships, work experience and voluntary opportunities

www.ibexearth.com/about

FOR UPDATED AND ADDITIONAL RESOURCES, INCLUDING EXTRA CHAPTERS, GO TO WWW.ENV.CAREERS

References

Beattie, Andrew (2019) 'The 3 Pillars of Corporate Sustainability' www.investopedia.com/articles/investing/100515/three-pillars-corporate-sustainability.asp (accessed 29 August 2020).

CSO Magazine (2020) 'Top 10 green companies' www.csomagazine.com/top10/top-10-green-companies (accessed 6 April 2021).

Chan, Emily, (n.d.) The Muse 'Green Gigs: How to Launch a Stellar Sustainability Career' www.themuse.com/advice/green-gigs-how-to-launch-a-stellar-sustainability-career (accessed 6 April 2021).

Corporate Knights (2019) '2019 Global 100' of the Most Sustainable Corporations in the World January 22, 2019 www.corporateknights.com/reports/2019-global-100/2019-global-100-results-15481153 (accessed 6 April 2021).

Deloitte (2021) 'Sustainability and Corporate Social Responsibility (CSR)' www2.deloitte.com/ru/en/pages/risk/solutions/sustainability-and-csr.html (accessed 6 April 2021).

European Union Press Release (2019) 'What is carbon neutrality and how can it be achieved by 2050?' www.europarl.europa.eu/news/en/headlines/society/20190926STO62270/what-is-carbon-neutrality-and-how-can-it-be-achieved-by-2050 (accessed 6 April 2021).

Gallo, Iciar (2020) 'Evolution of ISO 14001: The history of the leading environmental management standard' www.advisera.com/14001academy/blog/2020/01/21/history-of-iso-14001-why-is-it-so-popular (accessed 29 August 2020).

Green Business Network (2019) 'What's a Green Business?' www.greenbusinessnetwork.org/about/whats-a-green-business/ (accessed 6 April 2021).

IEMA (2017) 'IEMA Reports 8% Growth in Global ISO 14001 Data' www.iema.net/resources/news/2017/09/26/iema-reports-8-growth-in-global-iso-14001-data (accessed 29 August 2020).

Newsweek Green Ranking 2017 www.newsweek.com/top-10-global-companies-green-rankings-2017-18 (accessed 29 August 2020).

Study.com (2019) 'Environmental Sustainability Career Information and Options' www.study.com/articles/Environmental_Sustainability_Career_Information_and_Options.html (accessed 29 August 2020).

The 2020 2020 GlobeScan/SustainAbility Leaders Survey Report www.globescan.com/wp-content/uploads/2020/08/GlobeScan-SustainAbility-Leaders-Survey-2020-Report.pdf (accessed 6 April 2021).

Waite, M. (2013). SURF Framework for a Sustainable Economy. *Journal of Management and Sustainability* 3: 25–40. https://doi.org/10.5539/jms.v3n4p25.

8

Air Quality

8.1 Sector Outline

The National Institute of Environmental Health Sciences (2016) offer a definition of air quality:

> Typically, air quality is defined as the degree to which to air in a particular region is free from pollution. Good air quality tends to have pollution levels below those which are considered to be detrimental to human and ecological life. Clean air is necessary to support the delicate balance of life on Earth; this includes the world's oceans, forests, soils and all forms of existence. Poor air quality can occur through both natural and anthropogenic causes; this can include volcanic eruptions and thawing of permafrost, or industrialization and urbanization.

Conserve Energy Future (2009) highlights key causes of poor air quality:

- Burning of Fossil Fuels: The combustion of finite resources such as petroleum and coal is a primary cause of air pollution and poor air quality. Much of the developed world depends on burning fossil fuels for their basic needs in terms of transportation and in order to develop economically. This process, however, releases harmful greenhouse gases such as sulfur dioxide and carbon monoxide (from incomplete combustion)
- Agricultural activities: One of the most perilous gases in the atmosphere is ammonia, which also happens to be a very common product of agriculture activities
- Factories and Industries: Industries and factories rely heavily on the burning of non-renewable resources, which are responsible for the release of pollutants such as carbon monoxide and hydrocarbons, both of which deplete the quality of air
- Mining: This is a process whereby minerals are extracted from deep within the earth, often with the use of large equipment that in itself can be responsible for the destruction of large areas of natural ecosystems. In addition to this, however, during the process of extracting these minerals, chemical and dust are released also contribution to pollution. This has also been blamed for negative health effects of those working in this sector, in the form of respiratory diseases and cardiovascular diseases

Global Environmental Careers: The Worldwide Green Jobs Resource, First Edition. Justin Taberham.
© 2022 John Wiley & Sons Ltd. Published 2022 by John Wiley & Sons Ltd.

- Indoor Air Pollution: This can be in the form of residual pollutants such as asbestos, formaldehyde and lead from building materials, household cleaning products, particularly from paint, mold and tobacco smoke

The key effects of poor air quality include climate change, respiratory and cardiovascular diseases, eutrophication, acid rain, ecological damage and depletion of the earth's ozone layer.

8.2   Issues and Trends

The WHO (2018) states:

> Air pollution levels remain dangerously high in many parts of the world. New data from WHO shows that 9 out of 10 people breathe air containing high levels of pollutants. Updated estimations reveal an alarming death toll of 7 million people every year caused by ambient (outdoor) and household air pollution. . .More than 90% of air pollution-related deaths occur in low- and middle-income countries, mainly in Asia and Africa, followed by low- and middle-income countries of the Eastern Mediterranean region, Europe and the Americas. . . More than 4300 cities in 108 countries are now included in WHO's ambient air quality database, making this the world's most comprehensive database on ambient air pollution. Since 2016, more than 1000 additional cities have been added to WHO's database which shows that more countries are measuring and taking action to reduce air pollution than ever before.

Europe

The European Environment Agency reported in 2017:

> Air pollution harms human health and the environment. In Europe, emissions of many air pollutants have decreased substantially over the past decades, resulting in improved air quality across the region. However, air pollutant concentrations are still too high, and air quality problems persist. A significant proportion of Europe's population live in areas, especially cities, where exceedances of air quality standards occur: ozone, nitrogen dioxide and particulate matter (PM) pollution pose serious health risks. Several countries have exceeded one or more of their 2010 emission limits for four important air pollutants. Reducing air pollution therefore remains important.

Asia

AirQualityAsia (AQA 2021) highlighted key issues in Asian air quality:

> Asia's rapidly growing economies are producing a rising proportion of the world's global CO_2 emissions. . .The energy transition of Asia, in this critical window of the next three years, is also essential for global health as air pollution crosses boundaries, oceans, destroys human health, creates welfare costs to national GDP, and reduces agricultural output.

The Climate and Clean Air Coalition (CCAC) noted:

> The impact of air pollution on human health constitutes a serious public health crisis across Asia and the Pacific. About 4 billion people, around 92 per cent of the region's population, are exposed to levels of air pollution that pose a significant risk to their health: exposure to pollution levels in excess of the World Health Organization (WHO) Guideline for public health protection is associated with elevated risks of premature death and a wide range of illnesses. Reducing this health burden requires further action in Asia and the Pacific to reduce emissions that lead to the formation of fine particulate matter (PM2.5) and ground-level ozone, which undermine people's health and well-being as well as food production and the environment.

Air pollution in Asia can reach dangerous levels where the cities are covered by a blanket of smog, which is a great threat to human (Ives 2015).

Africa

The OECD stated in 2017:

> Air pollution is of significant and increasing concern for the continent. . .Between 1990 to 2013, total annual deaths from ambient particulate matter pollution (APMP), mostly caused by road transport, power generation or industry) rose by 36% to around 250 000, while deaths from household air pollution (HAP), caused by polluting forms of domestic energy use) rose by 18%, from a higher base, to well over 450 000. For Africa as a whole. . .the economic cost of premature deaths caused by each of these sources of pollution surpasses those associated with unsafe sanitation or underweight children.

An article by Kuo (2015) noted:

> Air pollution from African cities will increase dramatically over the next few decades. They warn that by 2030 Africa's anthropogenic emissions will be equal to those from forest fires due to a rise in vehicle use and industry in the region. It has also been suggested that Africa's carbon emissions could be responsible for almost 50% of the global total.

Jane Akumu of UNEP highlighted issues of concern in 2014:

> Air quality monitoring in Africa is set to become ever more important as it is predicted to be home to two fifths of the world's population by 2050. Furthermore, it has been suggested that as much as 94% of Nigeria's population are exposed to air pollution levels above what the WHO classify as safe.

North America

Since the passage of the Clean Air Act in 1970, American air quality has steadily improved. Emissions of air pollutants have decreased by around 1–3% each year, which may not seem like much, but over a 40-year period it has led to a decrease in harmful emissions of more than 50% (Greenfield 2011).

UNEP, in The North America Air Quality Regional Report, noted:

> Information from air quality monitoring stations across the sub-region indicates that air quality has generally improved over the last few decades; however, it still remains an issue of concern, causing approximately 44,000 premature deaths annually. In addition, specific exceedances of legal or recommended values in certain places still occur, especially for particulate matter (during wintertime) and ozone (in summer months).
>
> In Canada, the federal government sets the ambient air quality objectives and standards in conjunction with the provinces, while the provincial governments apply these standards through a wide range of environmental management tools. The World Health Organisation (WHO) estimates that outdoor air pollution causes approximately 2,700 premature deaths annually; however, a study by the OECD reviewed this number upwards to 7,469 in 2010.
>
> In the United States, air quality has greatly improved in the last few decades due to regulations, technology improvements and economic changes. . .However, approximately 57million people still live in areas with unhealthy levels of air pollution. Topography and weather conditions are some of the external factors that aggravate air pollution in the United States, especially in urban centres. . .The main sources of air pollution in the region are industries, heating of homes with biomass (particularly wood burning stoves and boilers), and transport (which is often the main source of urban air pollution). . .other anthropogenic sources of air pollution have considerable impact on human health. This category includes agriculture which is a major source of direct air pollution and also a source of precursor gases that after undergoing atmospheric processing become air pollutants with potential to impair human health.
>
> *Source: UNEP 'The North America Air Quality Regional Report'. Retrieve from www. unenvironment.org/resources/report/north-america-air-quality-regional-report.*

South America

Maxwell (2013) reported that:

> According to [a] new report, published by the Clean Air Institute (CAI), over 100 million people in Latin America breathe polluted air. The authors looked at levels of particulate matter (PM10 and PM2.5), ozone (O3), nitrous oxide (NO2), and sulfur dioxide (SO2) in the region. They compared the levels of those compounds with the World Health Organization (WHO)'s Air Quality Guidelines, and found that:

Of the 16 countries that measured for PM10 in 2011, all exceeded the WHO's recommended level

Of the 11 countries that measured for PM2.5 in 2011, 10 exceeded the WHO's recommended level

Ozone was difficult to measure, but the authors did manage to take measurements of ozone in 2011 in Santiago (Chile), Mexico City and Quito, and all three cities exceeded the WHO's recommended level. Of the 13 countries that measured for NO2, 7 exceeded the WHO's recommended level

Encouragingly, the researchers also found that many countries have some standards in place to limit these emissions:

Approximately half of the countries included in the study have PM2.5 standards
All of the countries that have any standards in place (16) have PM10 standards
13 of the countries have Ozone (8 hour) standards
All 16 countries have SO2 (24 hour) standards
15 of the 16 countries have (annual) NO2 standards

Source: Maxwell (2013). © Natural Resources Defense Council.

Oceania

According to the Australian Government (2015):

Australia's Environment Ministers established the National Clean Air Agreement on 15th December 2015. When compared to the rest of the world, Australia is doing relatively well in terms of its air quality. There are a number of strategies currently in place in Australia which work to monitor and reduce the levels of pollutants in the air. Regardless of this, ground level ozone and particulate matter still exceeds current air quality standards. There are also additional pressures, with respect to air quality, from population growth, rapid urbanization an increasing demand for energy and transportation fueling a more consumer driven lifestyle. In 1998, Australia's National Environment Protection Ambient Air Quality) Measure was established. This provided motivation to protect ecosystems and humans from the adverse effects of air pollution.

In the StatsNZ (2018) release 'Report shows New Zealand air quality is good' the following issues were highlighted:

while some previously known issues persist, progress has been made and levels of some pollutants are declining. . .Burning wood and coal for home heating in winter is the single leading cause of human-generated poor air quality. . .Vehicle emissions are also an important cause of human-generated poor air quality.

8.3   Key Organisations and Employers

Within the sector, key employers globally include consultants and central and local government.

Key organisations include:

Environmental Protection UK Air Quality Committee
www.environmental-protection.org.uk/policy-areas/air-quality

European Commission – European Thematic Strategy on Air Pollution
www.eea.europa.eu/policy-documents/thematic-strategy-on-air-pollution
www.environmental-protection.org.uk/policy-areas/air-quality/air-pollution-law-and-policy/european-air-quality-laws

North America – the following list is from Insteading.com (2016):

National Oceanic and Atmospheric Administration
www.noaa.gov

National Academies Board on Atmospheric Sciences and Climate
www.nationalacademies.org/basc/board-on-atmospheric-sciences-and-climate

USDA Agricultural Air Quality Task Force
www.airquality.nrcs.usda.gov

Environmental Protection Agency
www.epa.gov

Centers for Disease Control and Prevention
www.cdc.gov

Clean Air World
www.cleanairworld.org

National Association of Clean Air Agencies
www.4cleanair.org

Center for Clean Air Policy
www.ccap.org

World Meteorological Organization (WMO)
www.wmo.int

Air & Waste Management Association
www.awma.org

Many NGOs have programmes in Asia, including Greenpeace, who are currently running operations in mainland China to attempt to reduce the levels of air pollution in its major cities (Greenpeace East Asia 2015). In addition, the Asia Centre for Air Pollution Research (ACAP) works on The Acid Deposition Monitoring Network in East Asia (EANET) and researches the causes and effects of air pollution in East Asia comprising acid deposition and oxidant (Asia Centre for Air Pollution Research 2019).

There are a number of organizations who operate in Africa to monitor air quality in the region. One such operation is the South African Air Quality Information System, which is a database for managing air quality information in South Africa and makes data available to stakeholders to be analysed and managed.

8.4 Careers in the Sector

Air Pollution Analysts and Air Quality Field Technicians work to measure, sample and analyse data collected from polluted air. Their work often involves them discovering which pollutants are present in a sample and the source of those pollutants, solutions and theories can then be created to lower such levels of pollution or to place limitations on human pollution outputs. Working as an air pollution analyst, you can expect to spend a lot of time working in the field or in labs collecting and analysing data and generating possible strategies to lower or maintain such levels. A large majority of people working in this sector are employed by the government agencies or independent companies.

Air Pollution Analysts are often employed by government agencies on a federal, state or local level. The data collected often inspire environmental policy changes. However, private corporations and business will sometimes hire Air Pollution Analysts to determine the environmental detriments of their own procedures and practices (Environmental Science 2016).

Increasingly, environmental consultancies employ air quality specialists.

8.5 Job Titles in the Sector

Air pollution analyst, Air quality consultant, Senior Audit/EHS Consultant, Air Quality Planner, Air Monitoring & QA Specialist, Air Quality Specialist/Engineer, Air Compliance Specialist/Environmental Specialist, Postdoctoral Research Fellow, Agricultural Air Quality Modelling, Environmental Engineer – Air Quality.

8.6 Educational Requirements

- Science or mathematics-based undergraduate degree
- Often a higher-level degree is asked for by employers, but work experience may substitute for this
- Relevant work experience or an internship is valuable
- Some roles ask for environmental engineering qualifications

8.7 Personal Attributes and Skill Sets

There are many skills sets and attributes for air quality professionals, including:

- Project management
- Teamwork and the ability to work alone and remotely
- Excellent communication skills including the ability to get technical information across to different audiences such as committees and local groups
- Knowledge of current legislation (national and local) and trends in the sector
- Knowledge of air quality assessment techniques, regulatory requirements and EIA
- Understanding of industry processes
- Understanding of investigation and reporting

- Data analysis and technical reporting
- Ability to undertake fieldwork
- Higher level IT skills, possibly including GIS, mapping and modelling

Training

World Bank Open Learning Campus 'Introduction to Air Quality Management (Self-Paced)'
www.olc.worldbank.org/content/introduction-air-quality-management-self-paced

United States Environmental Protection Agency (EPA)
Air Quality Systems (AQS) Training
www.epa.gov/aqs/aqs-training

Clean Air Asia
www.cleanairinitiative.org/portal/knowledgebase/trainingcourses

European Commission 'Training Course on Air Quality and Health: Methods, Tools and Practices for Better Air Quality Action Planning'
www.ec.europa.eu/futurium/en/air-quality/training-course-air-quality-and-health-methods-tools-and-practices-better-air-quality

Clean Air Society of Australia and New Zealand (CASANZ)
www.casanz.org.au/course

National Association for Clean Air (NACA)
www.naca.org.za/courses.php

8.8 Career Paths and Case Studies

◄ Top Tips

Alun McIntyre, Technical Director and Air Quality Practice Lead at HaskoningDHV UK Ltd, UK

1) Before attending an interview, do your homework on the company, find out what it is good at and who the prime movers are within the company.
2) Do some wider background reading about the big emerging environmental issues – you need to know the landscape in which your chosen pathway resides.
3) Don't be afraid to dive carefully into new subject areas; I don't mean 'do something every day that scares you', just be accepting of wider opportunities, you may find something wonderfully satisfying disguised as something ordinary – that is the wonder of the natural environment.
4) Try not to fall out with all your peers, they will be there with you for your entire working life, it's a small professional community and you need to be appreciated.
5) Finally, listening is the greatest skill you can apply in understanding your clients' hopes, fears, challenges and real needs, therein lies your key to providing innovative solutions for them.

Source: Alun McIntyre, Technical Director and Air Quality Practice Lead at HaskoningDHV UK Ltd, UK. © John Wiley & Sons.

◀ **Top Tips**

Gill Cotter, Senior Air Quality Consultant, Mott MacDonald, UK

Air quality consultancy involves a wide range of tasks on a daily basis, with various assessment types, air dispersion modelling, air quality monitoring, project management, client management and providing advice to clients and colleagues.

   Career progression usually starts as an Air Quality Graduate Consultant, moving up to Consultant (approximately 2 years' experience), Senior Consultant (5 years), Principal Consultant (8 years) and then potentially Technical Director or equivalent (10+ years). Some consultants choose to work on a freelance basis once they are at a senior level and often work alongside companies as an Associate.

1) Get a Science or Mathematics-based Undergraduate Degree
   There's a lot of variation in air quality consultants' university degrees, but the majority are science-based, ranging from environmental studies, biology, physics and the more obvious atmospheric chemistry. A good level of numeracy is important when undertaking air quality assessment calculations.
2) Become Involved in Professional Bodies
   In the United Kingdom, this is the Institute of Air Quality Management (IAQM)/Institute of Environmental Sciences (IES). Joining the IAQM and IES as an Associate Member will give access to seminars and events that can help broaden your knowledge, whilst giving you the opportunity to network with other air quality professionals.
3) Be Familiar with Relevant Air Quality Policy and Guidance
   Air quality policy and guidance are referred to and utilised on a daily basis; therefore, a broad understanding of how these fit together when assessing air quality is important.
4) Learn about Geographical information Systems (GIS)
   This is a very useful tool for air quality modelling and creating maps for reports and can be learned free through opensource software such as QGIS. Including this on a CV can really boost appeal to employers.
5) Develop the Ability to Communicate Effectively with Clients and Colleagues
   Consultants have discussions with clients and colleagues in a daily basis and provide clear and concise written reports. Good communication skills are vital to ensure that you are giving your client what they need to meet their goals (not necessarily what they ask for). It is also important that you are able to relay complex information to clients so they can understand the outcome of assessments.

Source: Gill Cotter, Senior Air Quality Consultant, Mott MacDonald, UK. © John Wiley & Sons.

8.9   External Resources

UK

IAQM Institute of Air Quality Management
www.iaqm.co.uk

Environmental Protection UK Air Quality Committee
www.environmental-protection.org.uk/policy-areas/air-quality

Department for Environment Food and Rural Affairs UK AIR (Air Information Resource)
www.uk-air.defra.gov.uk/air-pollution

Europe

European Commission – European Thematic Strategy on Air Pollution
www.eea.europa.eu/policy-documents/thematic-strategy-on-air-pollution

The World Air Quality Project 'Air Pollution in Europe: Real-time Air Quality Index Visual Map'
www.aqicn.org/map/europe

European Environment Agency 'European Air Quality Index'
www.airindex.eea.europa.eu/Map/AQI/Viewer

World Health Organisation Regional Office for Europe 'Air Quality Data and Statistics'
www.euro.who.int/en/health-topics/environment-and-health/air-quality/data-and-statistics

Asia

Asia Centre for Air Pollution Research
www.acap.asia/en

Air Quality Asia
www.airqualityasia.org

The Acid Deposition Monitoring Network in East Asia (EANET)
www.eanet.asia

Africa

National Association for Clean Air (NACA)
www.naca.org.za

South African Air Quality Information System
www.saaqis.environment.gov.za

Yat Ho Yiu Earth.org (2019) 'Air Pollution Is Starting to Choke Africa'
www.earth.org/air-pollution-is-starting-to-choke-africa

North America

AirNow
www.airnow.gov/about-airnow

U.S. Environmental Protection Agency (EPA) Our Nation's Air
www.gispub.epa.gov/air/trendsreport/2019/#home

UNEP (2015) 'Air Quality Policies in United States of America'
www.unenvironment.org/resources/policy-and-strategy/air-quality-policies-united-states-america

Government of Canada – Air Quality
www.weather.gc.ca/mainmenu/airquality_menu_e.html

South America

UNEP (2019) 'Latin American and Caribbean communities mobilize to beat air pollution'
www.unenvironment.org/news-and-stories/press-release/latin-american-and-caribbean-communities-mobilize-beat-air-pollution

LEDS LAC 'Air quality, climate change and development in Latin America: opportunities for reducing short-lived climate pollutants'
www.ledslac.org/en/2016/06/air-quality-climate-change-and-development-in-latin-america-opportunities-for-reducing-short-lived-climate-pollutants

World Health Organisation (WHO) (1963) 'Atmospheric pollution in Latin America: interregional symposium on criteria for air quality and methods of measurement, Geneva, 6–12 August 1963'
www.apps.who.int/iris/handle/10665/326437

(A historical perspective)

Oceania

Clean Air Society of Australia and New Zealand (CASANZ)
www.casanz.org.au

Australian Government Bureau of Meteorology 'Smoke and air quality information'
www.bom.gov.au/catalogue/warnings/air-pollution.shtml

Environmental Justice Australia 'Reducing the health and environmental burden of coal-fired power'
www.envirojustice.org.au/our-work/community/air-pollution

Ministry for the Environment 'Monitoring air quality'
www.mfe.govt.nz/air/state-of-our-air/monitoring-air-quality

Global

Our World in Data (2019) Air Pollution
www.ourworldindata.org/air-pollution

World Health Organisation (WHO) 'Air Pollution'
www.who.int/health-topics/air-pollution

Christina Nunez, National Geographic (2019) 'Air pollution, explained'
www.nationalgeographic.com/environment/global-warming/pollution

FOR UPDATED AND ADDITIONAL RESOURCES, INCLUDING EXTRA CHAPTERS, GO TO WWW.ENV.CAREERS

References

AirQualityAsia (AQA) (2021). https://www.airqualityasia.org/about/ (accessed 6 April 2021).

Akumu, J. (2014). Improving air quality in African cities. www.wedocs.unep.org/handle/20.500.11822/16824 (accessed 09 February 2016).

Asia Centre for Air Pollution Research (2019). About ACAP. www.acap.asia (accessed 10 February 2016).

Australian Government (2015). Air quality. www.environment.gov.au/protection/air-quality (accessed 08 February 2016.

Climate and Clean Air Coalition (CCAC)/UNEP (2019). Air pollution in Asia and the Pacific: science-based solutions. https://www.ccacoalition.org/en/resources/air-pollution-asia-and-pacific-science-based-solutions-summary-full-report (accessed 6 April 9, 2021).

Conserve Energy Future (2009). Causes, effects and solutions of air pollution. www.conserve-energy-future.com/causes-effects-solutions-of-air-pollution.php (accessed 07 February 2016).

Environmental Science.org (2016). What is an air pollution analyst? www.environmentalscience.org/career/air-pollution-analyst (accessed 10 February 2016).

European Environment Agency (2017). Air pollution. www.eea.europa.eu/themes/air/intro (accessed 10 September 2020).

Greenfield, R. (2011). America's 10 best and 10 worst cities for air pollution. www.theatlantic.com/health/archive/2011/04/americas-10-best-and-10-worst-cities-for-air-pollution/237974 (accessed 09 February 2016).

Greenpeace East Asia (2015). Work for greenpeace. www.greenpeace.org/eastasia/about/jobs (accessed 11 February 2016).

Insteading.com (2016). Air Pollution Organizations. www.insteading.com/blog/organizations-air/ (accessed 10 September 2020).

Ives, M. (2015). The air pollution that's choking Asia. http://edition.cnn.com/2015/01/27/asia/asia-air-pollution-haze (accessed 11 February 2016).

Kuo, L. (2015). West Africa's air pollution is reaching dangerously high levels—and we don't know the worst of it. www.qz.com/487128/west-africas-air-pollution-is-reaching-dangerously-high-levels-and-we-dont-know-the-worst-of-it (accessed 10 February 2016).

Maxwell A. (2013). Air quality in Latin America: high levels of pollution require strong government action. www.nrdc.org/experts/amanda-maxwell/air-quality-latin-america-high-levels-pollution-require-strong-government (accessed 11 February 2016).

National Institute of Environmental Health Sciences (2016). Air pollution. www.niehs.nih.gov/health/topics/agents/air-pollution/ (accessed 10 February 2016).

Rana Roy (2016). The cost of air pollution in Africa. OECD Development Centre working paper no. 333. www.un.org/africarenewal/sites/www.un.org.africarenewal/files/The_cost_%20of_air%20pollution_in_%20Africa.pdf (accessed 6 April 2021).

StatsNZ (2018). Report shows New Zealand air quality is good. www.stats.govt.nz/news/report-shows-new-zealand-air-quality-is-good (accessed 6 April 2021).

UNEP. The North America Air Quality Regional Report. www.unenvironment.org/resources/
report/north-america-air-quality-regional-report (accessed 6 April 2021).

World Health Organisation (2018). 9 out of 10 people worldwide breathe polluted air, but more
countries are taking action. www.who.int/news-room/detail/02-05-2018-9-out-of-10-people-
worldwide-breathe-polluted-air-but-more-countries-are-taking-action (accessed 6
April 2021).

9

Fisheries Management

9.1   Sector Outline

Fisheries Management can be separated into two intertwined areas of the sector:

- Fisheries management (both freshwater and marine)
- Aquaculture and commercial fishing (freshwater and marine)

The UN FAO (1997) has defined fisheries management as:

> The integrated process of information gathering, analysis, planning, consultation, decision-making, allocation of resources and formulation and implementation, with enforcement as necessary, of regulations or rules which govern fisheries activities in order to ensure the continued productivity of the resources and the accomplishment of other fisheries objectives.
>
> From this description, it can be seen that fisheries management involves a complex and wide-ranging set of tasks, which collectively have the achievement of sustained optimal benefits from the resources as the underlying goal.

The European Commission (1995–2021) also notes, from a commercial fishing/aquaculture viewpoint:

> Fisheries management is based on data and scientific advice, and control measures to ensure that rules are applied fairly to and complied with by all fishermen. We all depend on healthy ecosystems: for food, energy, raw materials, air and water.

These helpful definitions focus on key themes in the sector:

- Integrated management
- Information and data gathering
- Data analysis
- Planning and consultation
- Decision making
- Resource management

- Regulation and rules, enforcement
- Continued productivity of resources

In all these themes, there are employment opportunities.

An excellent source document is The UN FAO Fisheries (2021) and Aquaculture Department's 'State of World Fisheries and Aquaculture' (SOFIA) which is an advocacy document published every two years.

There is considerable global evidence that where fish stocks are actively managed, fish populations are sustainable whereas in areas where fisheries management activity is less comprehensive, fish stocks tend to be poorer. Hilborn et al. (2020) gave global insight into this issue:

> Marine fish stocks are an important part of the world food system and are particularly important for many of the poorest people of the world. Most existing analyses suggest overfishing is increasing, and there is widespread concern that fish stocks are decreasing throughout most of the world. We assembled trends in abundance and harvest rate of stocks that are scientifically assessed, constituting half of the reported global marine fish catch. For these stocks, on average, abundance is increasing and is at proposed target levels. Compared with regions that are intensively managed, regions with less-developed fisheries management have, on average, 3-fold greater harvest rates and half the abundance as assessed stocks. Available evidence suggests that the regions without assessments of abundance have little fisheries management, and stocks are in poor shape. Increased application of area-appropriate fisheries science recommendations and management tools are still needed for sustaining fisheries in places where they are lacking. . .
>
> The efforts of the thousands of managers, scientists, fishers and nongovernmental organisation workers have resulted in significantly improved statuses of fisheries in much of the developed world, and increasingly in the developing world. Scientifically managed and assessed fish stocks in many places are increasing or are already at or above the levels that will provide a sustainable long-term catch. The major challenge now is to bring fisheries science methods and sustainability to fisheries that remain largely unassessed and unmanaged.

Recreational Fishing is an area which can have significant economic benefit but also implications for fisheries stock management and conservation. The FAO (2017) offers a definition:

> Recreational fishing (RF) is defined as the 'fishing of aquatic animals (mainly fish) that doesn't constitute the individual's primary resource to meet basic nutrition needs and are not generally sold or otherwise traded on export, domestic or black markets'.

The FAO also highlights a number of issues within RF:

- RF may harvest from the same stock as commercial fisheries
- RF multiplies the value of fish compared to commercial fishing and incomes from RF are high – The World Bank has estimated that anglers spend approximately US$190 billion

annually related to recreational fishing, contributing about USD$70 billion per year to global gross domestic product. These are probably low estimates, not including the large revenue streams for fishing tackle.

- The number of anglers globally could be from 220 to 700 million
- RF is increasing globally, especially in developed nations
- Some wild freshwater fish stocks as well as marine special are facing high exploitation rates
- Population profiles of some species can be altered
- Disturbance of habitats, target and other species
- The taking of 'trophy' fish
- Fishing pressure causing injury and disease among target species

There are measures to reduce impacts which include:

- Active fisheries management measures (often delivered by Government departments, government bodies, landowners and others)
- Licensing, regulation and enforcement
- Close seasons for fishing, especially in fish spawning seasons
- Catch limits and size limits

Rahel and Taniguchi (2019) compared freshwater fisheries management in the United States of America and Japan and noted:

> The USA and Japan differ in their approaches to managing inland recreational fisheries. The USA uses a public property rights regime whereby access rights are assigned to the states, which manage the fishery resource for the public good. Japan uses a common property rights regime whereby access rights for waterways are assigned to fishery unions, which manage the resource for the benefit of their members. Members of fishery unions are likely to develop an emotional attachment to the fishery that results in few regulation violations or illegal fish introductions. The USA would benefit from actions that promote such a caretaker attitude towards the environment. Habitat improvement is a major activity in the USA but is less prevalent in Japan where stocking is the dominant management activity. Catch-and-release angling, size restrictions and employment of professional fisheries biologists are more prevalent in the USA compared to Japan. The USA has a tax on fishing equipment that funds management activities whereas such a funding source is lacking in Japan. Despite differences in management regimes, both countries face similar challenges in recruiting new anglers and meeting the conflicting mandates to enhance sport fisheries while conserving native species.

Despite the impacts of RF, there are significant benefits in harnessing the fact that anglers are a powerful lobby and generally passionate about protecting the environment. Angling bodies are active in habitat management and with little government support achieve considerable benefit to the environment. The RF sector is key to the sustainable management of many aquatic environments, which should be taken into account at central government level.

There is growing employment in the recreational fisheries management sector.

9.2 Issues and Trends

A key trend globally is to bring active fisheries management to countries where this is currently patchy or weak in terms of regulations and enforcement. In this way, fisheries production and commercial fishing can be carried out in a more sustainable way.

The WWF (n.d.) highlight key challenges:

> In many cases, fisheries rules, regulations and enforcement measures are not efficient; fishing capacity and efforts are not sufficiently limited or controlled.
>
> Another important issue is that today's fishing activities often occur far from the eye of regulators and consumers. . .
>
> Inadequate fisheries regulations: In many fisheries, current rules and regulations are not strong enough to limit fishing capacity to a sustainable level
>
> Lack of implementation/enforcement: . . .many countries have still not ratified, implemented, or enforced international regulations such as the UN Convention on the Law of the Sea and the UN Fish Stocks Agreement.
>
> Lack of transparency and traceability: Customs agencies and also retailers are not always ensuring that the fish entering their country and shops is caught legally and in a sustainable way. As a result, consumers are unwittingly supporting poor management by purchasing fish from unsustainable fisheries.
>
> Failure to follow scientific advice: Many fisheries management bodies do not heed scientific advice on fish quotas and set catch limits above the recommended maximum amount
>
> Flag of Convenience vessels: Countries are either failing to restrict fishing companies from owning and operating FoC vessels or are not rigorously inspecting FoC vessels landing at their ports. This includes countries with some of the biggest fishing fleets. . . This allows illegal, unreported and unregulated (IUU) fishing to continue.
>
> Too few no-go areas for fishing: Protected areas and no-take zones, where fishing is banned or strictly regulated, can provide essential safe havens where young fish can grow to maturity and reproduce before they are caught. The current lack of protection is especially worrying for fish spawning grounds and the deep sea, both of which are particularly vulnerable to overfishing.

In Africa, the EAF (The ecosystem approach to fisheries) was a key objective of the EAF-Nansen Project which ran between 2006 and 2016. The objective of the programme was:

> to enable African coastal countries to manage their fisheries in a way that would safeguard the health of marine ecosystems.

Measures within the project aimed to implement fisheries management across 31 coastal countries in Africa. The project report (UN FAO 2016) notes:

> It was also a pioneering initiative for the African continent where environmental degradation was on the increase, there was limited awareness of the importance of controlling the impact of fisheries on marine ecosystems, and little scientific data on the conditions of marine ecosystems and the impact of human activity on them.

The main focus of the project has been on capacity development, with an emphasis on laying the groundwork for ownership and sustainability within partner countries. Its specific goal has been to enable nations to draft and implement their own fisheries management plans according to the principles of EAF, and to empower regional fisheries bodies to serve their member states as they begin implementing EAF.

The EAF-Nansen Project has laid important foundations for supporting a sustainable and unified ecosystem approach to fisheries management across Africa. In the long term this will contribute to food security, protection of resources and livelihoods and the greater health of marine ecosystems. While the focus has been on Africa, the knowledge and experience generated extends to the rest of the world, and in particular those developing countries that still lag behind the developed world in terms of ecosystem protection and sustainable fisheries management.

The second phase of the project started in 2017.

In addition, the World Bank (2021) developed its 'Africa Program for Fisheries':

The World Bank supports Africa's commitment to invest in sustainable fisheries as a way to build the resilience and improve the livelihoods of coastal communities. Since 2005, the Africa Program for Fisheries has focussed on sustainable use of the marine resources, governance of the sector and deep engagement with coastal communities. Transformative interventions are giving new hope to coastal communities.

Fisheries play significant social and nutritional roles in Africa. The sector contributes to food and nutrition security, and provides jobs, in particular for coastal populations, which are often among the poorest and most vulnerable. On average globally, fish and fish products account for 18% of animal protein intake. Due to the growing population and per capita income, demand for fish is expected to increase 30% by 2030. If the current trend continues without management, the poorest countries will suffer the most. Climate change aggravate these challenges with rising sea temperatures, harsher weather conditions for fishers, migration of fish to cooler waters away from the equator and shrinking fish size.

Fisheries contribute to Africa's economy. Currently, fisheries and aquaculture directly contribute $24 billion to the African economy, representing 1.3% of the total African GDP in 2011. The sector provides employment to over 12 million people (58% in the fishing and 42% in the processing sector). . .

In 2005, the Bank created a worldwide program to promote and facilitate fisheries and aquaculture's contribution to sustainable growth, food security, women empowerment and poverty reduction. The Global Program on Fisheries (PROFISH) provided information, knowledge products and expertise to help design and implement good governance.

In Latin America, the FAO noted:

It is estimated that fisheries and aquaculture provide a livelihood for 540 million people worldwide, 8% of the world's population. The seas of Latin America and the Caribbean are a source of healthy food and a resource for thousands of families. According to FAO data, in 2009 aquaculture provided 81% of the seafood, 76% of the freshwater fish, 69% of the salmon and 42% of the shrimp consumed in the world, generating employment for 9 million people. Aquaculture is also the world's fastest-growing food industry: 7% per year, accounting for more than 50% of fish for human consumption.

The contribution of aquaculture to the regional economy has grown substantially in the last 10 years. It provides employment for more than 200,000 people directly and approximately 500,000 indirectly. From a food perspective, more than 100,000 rural families in the region depend directly or indirectly on aquaculture for their livelihood, including food for private consumption.

Source: UN Food and Agriculture Organisation (UN FAO) (2020).

The Asia Foundation (2018) reported some challenges in fisheries management in Asia:

Approximately 12 percent of the world's population relies upon fisheries and aquaculture for their livelihood, and over half of the world's people get a significant source of their animal protein from fish and seafood. In Southeast Asia, this proportion is significantly higher. The region's seas not only serve as a major source of food and livelihood for hundreds of millions of people, they generate several billion dollars in GDP for the region.

Southeast Asia has one of the most diverse marine ecosystems in the world, but overfishing and destructive fishing threaten its sustained existence. Across the region, 64 percent of the fisheries' resource base is at a medium to high risk from overfishing, with Cambodia and the Philippines among the most heavily affected.

Much of the overfishing and destructive fishing in Southeast Asia is attributable to illegal, unreported and unregulated fishing (IUU). IUU fishing occurs region-wide, with violators ranging from small-scale local fishermen to large-scale enterprises conducted on commercial fishing trawlers. There are many drivers for IUU fishing in the region, not the least of which is that demand now appears to exceed supply. Operationally, the main issue is weak fishing regulations among the region's many countries, together with a lack of cooperation on management among these countries.

There is also a significant lack of science-based knowledge about the region's marine ecosystems to inform policies that would lead to the establishment of sound models for fisheries management, as well as insufficient focus on cultivating alternatives to wild catch fisheries, such as sea-farming and inland freshwater aquaculture.

Source: Kim J. DeRidder and Santi Nindang, Southeast Asia's Fisheries Near Collapse from Overfishing, The Asia Foundation, March 28, 2018.

9.3   Key Organisations and Employers

There are many key organisations in the sector at Federal/National level as well as Regulators such as the UK Environment Agency and a growing consultancy sector, who often have an involvement in Government fisheries activities. The UN is actively involved in the sector, as are funding bodies such as the World Bank and African Development Bank.

United Kingdom

The United Kingdom's changing position in Europe has amended its fisheries policies. POST (2018) notes:

> Following EU withdrawal the UK will have full responsibility for fisheries policy and management within its waters. . . After EU and CFP [Common Fisheries Policy] withdrawal, the UK will become an independent coastal state with sovereign rights to govern its 200 nautical mile EEZ under the United Nations Convention on the Law of the Sea (UNCLOS).

Recreational fisheries in England are regulated by the Environment Agency.

North America

NOAA (National Oceanic and Atmospheric Administration) Fisheries is responsible for managing marine fisheries within the U.S. exclusive economic zone. NOAA is the parent body for the National Marine Fisheries Service
www.fisheries.noaa.gov/insight/understanding-fisheries-management-united-states

US Fish and Wildlife Service
The mission of the Service is to 'work with others to conserve, protect and enhance fish, wildlife and plants and their habitats for the continuing benefit of the American people'.
www.fws.gov

Fisheries and Oceans Canada
Fisheries and Oceans Canada is the federal lead for safeguarding our waters and managing Canada's fisheries, oceans and freshwater resources.
www.dfo-mpo.gc.ca/index-eng.htm

Australasia

The Australian Fisheries Management Authority (AFMA) is the Australian Government agency responsible for the efficient management and sustainable use of Commonwealth fish resources on behalf of the Australian community.
www.afma.gov.au

Fisheries New Zealand works to ensure that fisheries resources are managed to provide the greatest overall benefit to New Zealanders. Our focus is the sustainability of New Zealand's wild fish stocks, aquaculture and the wider aquatic environment, now and for future generations.
www.fisheries.govt.nz/fisheriesnz

There are many NGOs working in the fisheries management sector, both globally and nationally. These include:

Global Fishing Watch

Their mission is: 'We're committed to advancing ocean sustainability and stewardship through increasing transparency. We do this by offering, for free, data and near real-time tracking of global commercial fishing activity, supporting new science and research, and boosting the global dialogue on ocean transparency'.

www.globalfishingwatch.org/about-us

WorldFish

The WorldFish mission is to strengthen livelihoods and enhance food and nutrition security by improving fisheries and aquaculture. We pursue this through research partnerships focussed on helping those who stand to benefit the most – poor producers and consumers, women and children.

www.worldfishcenter.org

The Institute of Fisheries Management (IFM)

The IFM is an international organisation, dedicated to the advancement of sustainable fisheries management. Membership is open to anyone with an interest in fish and fisheries, their proper management and conservation.

www.ifm.org.uk

American Fisheries Society (AFS)

The American Fisheries Society is the world's oldest and largest organisation dedicated to strengthening the fisheries profession, advancing fisheries science and conserving fisheries resources.

www.fisheries.org

9.4 Careers

In terms of job seeking in fisheries, there are many jobs boards and educational resources in the sector.

These include:

American Fisheries Society Jobs
www.fisheries.org/employment/jobs

US Fish and Wildlife Service jobs:
www.usajobs.gov/Search/Results/?l=United%20States&a=IN15&s=startdate&p=1

USA Jobs
www.usajobs.gov/Search/ExploreOpportunities/?Series=0482

Canadian Aquaculture Industry Alliance
www.aquaculture.ca/careers-in-aquaculture-index

Australian Fisheries Management Authority (AFMA)
www.afma.gov.au/about/careers

Careers.gov.nz 'Aquaculture Farmer'
www.careers.govt.nz/jobs-database/farming-fishing-forestry-and-mining/aquaculture-fishing/aquaculture-farmer/job-opportunities

World Aquaculture Society jobs
www.was.org/wases/Jobs/index.aspx

Prospects, UK Fisheries Officer
www.prospects.ac.uk/job-profiles/fisheries-officer

Global Aquaculture jobs
www.worldfishing.net/jobs-board

Institute of Fisheries Management jobs
www.ifm.org.uk/job-vacancies

Target Jobs Fisheries Officer, United Kingdom
www.targetjobs.co.uk/careers-advice/job-descriptions/279535-fisheries-officer-job-
    description

Fisheries Biologist jobs, USA
www.indeed.com/q-Fishery-Biologist-jobs.html

Find a Fishing Boat – commercial fishing jobs, United Kingdom
www.findafishingboat.com/crew-list/fishing-marine-jobs-crew-wanted

NOAA Fisheries jobs, USA
www.nmfs.noaa.gov/pr/about/jobs.htm
www.fisheries.noaa.gov/about-us/careers

Fisheries management and aquaculture careers
www.allaboutcareers.com/careers/career-path/fisheries-management-aquaculture

Fishery Manager role information and jobs
www.environmentalscience.org/career/fishery-manager

Fish Farmer Profile
www.nationalcareers.service.gov.uk/job-profiles/fish-farmer

Fish Farmer Magazine Jobs
www.fishfarmermagazine.com/jobs

Fisheries Officer Profile
www.prospects.ac.uk/job-profiles/fisheries-officer

UN Fisheries Jobs:
www.unjobs.org/themes/fisheries

World Fishing and Aquaculture Jobs:
www.worldfishing.net/jobs-board

Aquaculture Jobs:
www.hijobs.net/jobs/aquaculture

Aquaculture Talent
https://aquaculturetalent.com

9.5   Job Titles in the Sector

Fisheries Manager, Fisheries Scientist, Fish Farm Manager, Fisheries Researcher, Fisheries Officer, Fishery Assessment Officer, Fisheries Biologist, River Keeper, Bailiff, Hatchery Manager, Hatchery Technician, River Warden, Aquaculture Researcher, Aquaculturist, Deck Hand, Operative, Crew.

9.6   Educational Requirements

There are varying educational requirements in the sector:

- Manual roles in the sector do not need degree level qualifications
- Technical roles will need a degree level qualification in fisheries management, aquaculture or a natural sciences subject
- Research and academic roles often seek higher level degrees
- Specific sector training is an advantage
- Volunteering, internship or work experience is valuable

9.7   Personal Attributes and Skill Sets

Skills and personal attributes may include:

- Teamwork and interpersonal skills
- Communications skills
- Data analysis and reporting
- IT skills such as GIS, modelling and statistical analysis
- Knowledge of sector legislation, regulations and trends
- Specific species knowledge
- Community engagement skills
- Project management
- Ability to carry out fieldwork in all weathers

Training

American Fisheries Society – Continuing Education
www.fisheries.org/membership/continuing-education

IFM Training
https://ifm.org.uk/ifm-training

Sparsholt College
https://www.sparsholt.ac.uk/subject/fishery-studies-and-aquatics

Hadlow College
https://www.hadlow.ac.uk/courses/fisheries-management

Reaseheath College
www.reaseheath.ac.uk/further-education/courses/aquatic-ecosystems-fisheries-
  management

Shuttleworth College
https://www.bedford.ac.uk/our-courses/subjects/fisheries-and-fish-management

Seafish – Careers and Training in Aquaculture
www.seafish.org/safety-and-training/careers-in-the-uk-seafood-industry/career-paths/
  careers-and-training-in-aquaculture

FAO Fisheries Training resources:
www.fao.org/fishery/fishcode-stf/training/en
In almost all fisheries management roles there is a high level of on the job training.

9.8   Career Paths and Case Studies

Top Tips

Ian Dolben, Technical Advisor, Environment Agency, United Kingdom

1) Be flexible – especially in location. You may need to relocate several times to get on in an environmental career
2) Take every opportunity to enhance your skill set – especially in areas where certified abilities are essential; e.g. electrofishing in fisheries or crayfish handling
3) Keep up to date with research areas – good CPD shows commitment
4) Build networks – the environmental field is relatively small, and it is helpful to know who is doing what/where
5) Don't be time limited – environmental work is generally not 9–5 so you need to be prepared to work later or at weekends (sometimes for no pay!)

Source: Ian Dolben, Technical Advisor, Environment Agency, UK. © John Wiley & Sons.

Personal Profile

Sarah Hussey, Fisheries Biologist, Sea Farms Ltd

Working for a seafood company, I work on investigating our wild fishery supply chains. My role involves assessing the performance of fisheries against fishing, environmental and social compliance criteria, and supporting suppliers on implementing improvements such as Fishery Improvement Projects (FIPs) and meeting challenges such as independent certifications. We supply premium shellfish and seafood products predominately for the UK retail and foodservice sectors. I am lucky enough to travel all over the world as part of my role and can finally say, I have my dream job.

With over 10 years' experience of working in fisheries, I sit on a couple of advisory boards. One being the technical advisory board for the Responsible Fishing Scheme (RFS) version 2, now known as the Responsible Fishing Vessel Standard (RFVS) www.seafish.org/article/rfs-version-2.

I also sit on the Steering group for the Young Seafood Leaders Network (YSLN) – www.seafish.org/article/young-seafood-leaders-network.

As I represent the YSLN, I sit as part of the Seafood Leaders Industry Group (SILG) – www.seafish.org/article/seafood-industry-leadership-group-silg. This group leads and supports the Seafood 2040 Strategic Framework for England (SF2040) whilst overseeing and facilitating its 25 recommendations for the seafood industry.

Prior to working in seafood, I worked for the Environment Agency, the conservation sector and in private consultancy. Previous roles include Fisheries and Freshwater Business Manager for Thomson Ecology Ltd where I managed and developed the fisheries and aquatic part of the business. I enjoyed the combination of using my technical ability to advise clients, in addition to possessing the business acumen to build partnerships and strategically manoeuvre and position the company within the sector. I was proud to leave a legacy in the fisheries team on leaving the company for another opportunity I couldn't turn down.

I read Zoology and gained an Honours Degree (BSc Hons) at Queen Mary, University of London in 2006 and went on to gain a Master's degree (MSc) in Aquatic Resource Management (ARM) at King's College, University of London.

I feel very lucky to work in the field that I have such a passion for and studied. My love of fish began as a small child, when my Father would take myself, sister and brother fishing with him around our home area in North London. I also have fond memories of us visiting family in Bournemouth and going crabbing during the school holidays. My Zoology degree was water focussed and my career developed from there. Many a field trip was spent at Millport marine station.

To anyone wanting to pursue a career in the fisheries, aquatic or indeed seafood sector, I would say, you firstly need passion and a lot of determination and then some more! I volunteered part-time on the days I wasn't working or at University to gain some work experience, which is invaluable when you are starting your early career. Local wildlife groups are a good place to start and you can join professional institutions such as the Institute of Fisheries Management (IFM) which are great forums for networking. If you keep striving, you will find a way to succeed. I have also been very fortunate to meet some inspiring people along the way and have worked with some fantastic managers to look up to, including those from the Environment Agency and the Zoological Society of London (ZSL).

I started in a junior position at Thomson and worked my way up. I seized an opportunity where I saw a gap in the market for their business and began winning work and bringing new clients in the fisheries and aquatic sector. You certainly need drive and ambition when heading up a business division. Since 2014, I have been a committee member for the IFM for the London and South East branch. In late 2019, I was appointed a trustee of the Living River Foundation, which I feel honoured to be part of.

With over 10 years of experience, I feel I still have so much to learn. I want to continue learning and developing and would encourage those in the sector to keep inspiring the new generation of fisheries and aquatic ecologists.

www.linkedin.com/in/sarahhussey/

Source: Sarah Hussey, Fisheries Biologist, Sea Farms Ltd. © John Wiley & Sons.

Personal Profile

Nashon Amollo, Intern with Kenya Marine and Fisheries Research Institute (KMFRI), Kenya

My role entails research on fish population dynamics and general water ecology, including identification of commercially and ecologically important species, their distribution and stock assessment; collecting and disseminating scientific information on fisheries resources which will form the basis for their utilisation; studying and isolating suitable fish species for culture both in marine and freshwater; establishing a marine and freshwater collection to be used for research and training purposes; pollution monitoring; research and socio-economic studies.

As a researcher in KMFRI you have to have studied a course that covers multiple areas including biology, environmental science, analytical chemistry, microbiology and biotechnology, natural resources management, aquatic science, food science, sociology and economics. I graduated with a BSc in Natural Resource Management and also possess a Diploma in Wildlife Management. These qualifications were ideal for my role. It is my hope that through their new Strategic Plan, KMFRI will provide leadership and authority in informing sustainable management of the country's aquatic resources for national development.

Organisational Profile
Kenya Marine and Fisheries Research Institute (KMFRI)
KMFRI is a state corporate body, established in 1979 under the Science and Technology Act (Cap 250), which has since been repealed by the Science, Technology and Innovation Act No. 28 of 2013. KMFRI is under the Ministry of Agriculture Livestock and Fisheries.

KMFRI is mandated to: Undertake research in marine and freshwater fisheries, aquaculture, environmental and ecological studies and marine research including chemical and physical oceanography, so as to provide scientific data and information to enhance sustainable exploitation, management and conservation of Kenya's fisheries resources, and promote aquatic environmental protection, food security, poverty alleviation and employment creation;

KMFRI adopted a Programme approach in research based on six research programmes, i.e.

- Aquaculture Programme
- Environment and Ecology Programme
- Fisheries Programme

- Information and Database Programme
- Natural Products Programme
- Socio-economic Programme

One of their key projects is the South West Indian Ocean Fisheries Programme (SWIOFP) is an ambitious multinational research project with an overall goal that will see the West Indian Ocean's marine resources ecologically managed for sustainable use and benefit by the region's riparian countries. The project forms part of the Large Marine Ecosystem Programme approach (LME) and is supported by the Global Environment Facility (GEF) as a contribution to its international waters programme and implemented by the World Bank. Over the next five years, nine countries of the Western Indian Ocean will work together to understand and management better their fisheries through an LME (Large Marine Ecosystem) Based Approach.

As the leading aquatic research institution in Kenya, Kenya Marine and Fisheries Research Institute (KMFRI) has to deal with a broad range of issues which are very dynamic. Due to challenges facing the management of aquatic resources, KMFRI is now much more under pressure than ever before from the Government, the public and private sector to deliver relevant, practical and readily adaptable results for aquatic resource management.

The aquatic environment has continued to face diverse challenges including; over-exploitation of fish resources, pollution, habitat destruction and other forms of environmental degradation, conflicts in resource use and more recently, climate change. Therefore, KMFRI is expected to provide research solutions for sustainable exploitation of both freshwater and marine ecosystems. With the ever-increasing demand for fish as a result of rapid population growth, KMFRI must provide leadership in developing sustainable and economically viable aquaculture as an industry for employment creation, food security and income generation.

Source: Nashon Amollo, Intern with Kenya Marine and Fisheries Research Institute (KMFRI), Kenya. © John Wiley & Sons.

9.9   External Resources

United Kingdom

Institute of Fisheries Management, United Kingdom
www.ifm.org.uk

Fisheries Society of the British Isles (FSBI)
www.fsbi.org.uk

Marine Stewardship Council:
www.msc.org/about-the-msc/working-at-the-msc

University of Stirling Institute of Aquaculture
www.stir.ac.uk/about/faculties/natural-sciences/aquaculture

Europe

European Community Fisheries 'Employment'
www.ec.europa.eu/fisheries/3-employment_en

European Commission 'Managing fisheries'
www.ec.europa.eu/fisheries/cfp/fishing_rules_en

European Council 'Management of the EU's fish stocks'
www.consilium.europa.eu/en/policies/eu-fish-stocks

The Pew Charitable Trusts (2015) 'Timeline: Fisheries Management in North-Western European Waters'
https://www.pewtrusts.org/-/media/assets/2015/03/turningtide_infographic.pdf?la=de&hash=055F693708D2D37498D96FCAD0F995C7C816C3F7

Asia

Fisheries management in Asia-Pacific
www.fao.org/3/af347e03.htm

WorldFish 'Fisheries co-management in Asia: phase 1 project report'
www.worldfishcenter.org/content/fisheries-co-management-asia-phase-1-project-report

FAO Asia-Pacific Fishery Commission
www.fao.org/apfic/en

Africa

Farm Africa
www.farmafrica.org/agriculture/fisheries

The Fish Tank, blog article 'How aquaculture in Africa is benefiting from new technologies and best management practices'
www.blog.worldfishcenter.org/2019/01/aquaculture-in-africa-benefiting-from-new-technologies-best-management-practices/

Future Agricultures (2011) 'CAADP and Fisheries Policy in Africa: are we aiming for the right reform?'
www.assets.publishing.service.gov.uk/media/57a08adb40f0b652dd00093c/FAC_Policy_Brief_No40.pdf

FAO (2014) 'FAO Fisheries and Aquaculture Circular No. 1093 The Value of African Fisheries'
www.fao.org/3/a-i3917e.pdf

North America

American Fisheries Society
www.fisheries.org

Scholarships:
www.scholarshipdb.net/phd-in-fisheries-scholarships-in-United-States

Fisheries and Oceans Canada:
www.dfo-mpo.gc.ca/career-carriere/index-eng.html

NOAA Fisheries
www.nmfs.noaa.gov/stories/2015/11/msa40.html

NOAA Marine Aquaculture definition
www.nmfs.noaa.gov/aquaculture/what_is_aquaculture.html

South America

FAO Fisheries and Aquaculture Technical Paper. No. 544 'Coastal fisheries of Latin America and the Caribbean'
www.fao.org/3/i1926e/i1926e00.htm

Latin American Reduction Fisheries Supply Chain Roundtable (SR)
www.sustainablefish.org/Programs/Improving-Wild-Fisheries/Seafood-Sectors-Supply-Chain-Roundtables/Reduction-Fisheries/Latin-American-Reduction-Fisheries-SR

Konrad Adenauer Foundation and Sociedad Peruana de Derecho Ambiental (SPDA) (2019) 'Policy Brief on Trade and Environmental Policy - Small-Scale Fisheries in Latin America and the Caribbean: Sustainability Considerations'
www.kas.de/documents/273477/5442457/PB+N%C2%B07+Small-scale+fisheries+in+LAC_Eng.pdf

Oceania

OECD Country Note on Fisheries Management Systems – Australia
www.oecd.org/australia/34427707.pdf

Fisheries New Zealand
www.fisheries.govt.nz/fisheriesnz

Western Australia
www.fish.wa.gov.au

Northern Territory
www.fisheries.nt.gov.au

South Australia
www.pir.sa.gov.au

Queensland
www.dpi.qld.gov.au/fishweb

New South Wales
www.fisheries.nsw.gov.au

Victoria
www.dpi.vic.gov.au

Tasmania
www.dipwe.tas.gov.au

Global

WorldFish
www.worldfishcenter.org
www.worldfishcenter.org/publications-resources

FAO Fisheries Management overview
www.fao.org/docrep/005/y3427e/y3427e03.htm

FAO Aquaculture overview
www.fao.org/fishery/aquaculture/en

FAO Fisheries and Aquaculture Department 'The State of World Fisheries and Aquaculture' (SOFIA)
www.fao.org/fishery/sofia/en

OECD (2018) 'OECD Review of Fisheries 2017'
www.one.oecd.org/document/TAD/FI(2017)14/FINAL/en/pdf

UN Aquaculture:
www.unjobs.org/themes/aquaculture

World Aquaculture Society
www.was.org/wases/job/list

Worldfishing.net – Global Fisheries and Aquaculture Organisations
www.worldfishing.net/directory/categories/organisations

WWF Unsustainable Fishing
wwf.panda.org/our_work/our_focus/oceans_practice/problems/unsustainable_fishing/

Fish Information and Services
www.fis.com

Rare Fisheries pages
www.rare.org/program/fish-forever/

FOR UPDATED AND ADDITIONAL RESOURCES, INCLUDING EXTRA CHAPTERS, GO TO WWW.ENV.CAREERS

References

Asia Foundation (2018). Southeast Asia's fisheries near collapse from overfishing. www.asiafoundation.org/2018/03/28/southeast-asias-fisheries-near-collapse-overfishing/ (accessed 21 September 2020).

European Commission (1995–2021). Managing fisheries. www.ec.europa.eu/fisheries/cfp/fishing_rules_en (accessed 13 April 2021).

FAO (1997). Fisheries Management, FAO Technical Guidelines for Responsible Fisheries, vol. 4. Rome: FAO. 82p.

Hilborn, R., Amoroso, R.O., Anderson, C.M. et al. (2020). Effective fisheries management instrumental in improving fish stock status. *Proceedings of the National Academy of Sciences* 117 (4): 2218–2224. https://doi.org/10.1073/pnas.1909726116.

Rahel, F.J. and Taniguchi, Y. (2019). A comparison of freshwater fisheries management in the USA and Japan. *Fisheries Sciences* 85: 271–283. https://doi.org/10.1007/s12562-019-01291-6.

The Parliamentary Office of Science and Technology (POST) (2018). UK fisheries management. https://post.parliament.uk/research-briefings/post-pn-0572/ (accessed 13 April 2021).

UN FAO (2016). Refocusing fisheries management in Africa. www.fao.org/3/a-i6008e.pdf (accessed 13 April 2021).

UN FAO (2017). The role of recreational fisheries in the sustainable management of marine resources. GLOBEFISH highlights – issue 2 www.fao.org/in-action/globefish/fishery-information/resource-detail/zh/c/1013313/ (accessed 13 April 2021).

UN FAO Fisheries and Aquaculture Department (2021). The state of world fisheries and aquaculture (SOFIA). www.fao.org/fishery/sofia/en (accessed 13 April 2021).

UN Food and Agriculture Organisation (UN FAO) (2020). Fisheries and aquaculture production in Latin America and the Caribbean. www.fao.org/americas/prioridades/pesca-y-acuicultura/en/ (accessed 13 April 2021).

World Bank (2021). Africa program for fisheries. www.worldbank.org/en/programs/africa-program-for-fisheries (accessed 13 April 2021).

WWF (n.d.). Fishing problems: poor fisheries management. wwf.panda.org/our_work/our_focus/oceans_practice/problems/fisheries_management/ (accessed July 2020).

10

Marine Science and Conservation

10.1   Sector Outline

In recent years, marine science and oceanography seem to have become synonyms, although this issue is still up for debate. However, marine conservation has grown in scope but is still considered a sub-sector of marine science, as is the sub-sector of marine biology. It is helpful to explore the different terms within the sector and their definitions.

Educalingo.com (2020) brings several terms into its definition:

> 'Oceanography, also known as oceanology and marine science, is the branch of Earth science that studies the ocean. It covers a wide range of topics, including marine organisms and ecosystem dynamics; ocean currents, waves and geophysical fluid dynamics; plate tectonics and the geology of the sea floor; and fluxes of various chemical substances and physical properties within the ocean and across its boundaries. These diverse topics reflect multiple disciplines that oceanographers blend to further knowledge of the world ocean and understanding of processes within: astronomy, biology, chemistry, climatology, geography, geology, hydrology, meteorology and physics.'

Science Daily (2021) offers a definition for marine conservation:

> 'Marine conservation, also known as marine resources conservation, is the protection and preservation of ecosystems in oceans and seas. Marine conservation focuses on limiting human-caused damage to marine ecosystems, and on restoring damaged marine ecosystems. Marine conservation also focuses on preserving vulnerable marine species.'

The MarineBio Conservation Society (1998–2021) define marine biology as:

> marine biology is the study of life in the oceans and other saltwater environments such as estuaries and wetlands. . .The study of marine biology includes a wide variety of disciplines such as astronomy, biological oceanography, cellular biology, chemistry, ecology, geology, meteorology, molecular biology, physical oceanography and zoology and the new science of marine conservation biology draws on many longstanding scientific disciplines such as marine ecology, biogeography, zoology, botany, genetics, fisheries biology, anthropology, economics and law.

Global Environmental Careers: The Worldwide Green Jobs Resource, First Edition. Justin Taberham.
© 2022 John Wiley & Sons Ltd. Published 2022 by John Wiley & Sons Ltd.

National Geographic (2019) defines oceanography:

> Oceanography applies chemistry, geology, meteorology, biology and other branches of science to the study of the ocean. It is especially important today as climate change, pollution and other factors are threatening the ocean and its marine life.

Environmental science.org (2021), in their article 'What is an Oceanographer?' outline different elements within oceanography:

> An oceanographer is a. . .scientist who studies the ocean. . .
>
> Many discoveries made in the field of oceanography are the product of multidisciplinary and comprehensive efforts involving oceanographers from all from branches of the science.
>
> Marine biologists are oceanographers that study marine ecosystems and their inhabitants. This can involve working with research animals or taking trips into the ocean to perform different experiments, collect data, or track the animals.
>
> Physical oceanographers are more concerned with studying the movements of the oceans, in the waves and currents and tides that move the water itself.
>
> Chemical oceanographers monitor the chemical composition of the ocean water to better understand how they shape the planet. They may study pollution or help find naturally occurring resources on the seafloor.
>
> Geological oceanographers focus on studying the ocean's floor. They may study undersea volcanic activity and its relation to the movement of tectonic plates or the deep oceanic trenches.

There are opportunities to work in marine science across the globe as all countries with coastal waters have ecosystems, species and biodiversity requiring protection and management. Often these ecosystems contain habitats and species that are shared with adjacent countries meaning these countries need to work together to protect and manage them. Marine science involves assessing (surveys and monitoring), understanding (research and analysis), protecting, managing and restoring (conservation actions, resource and biodiversity management plans), the richness and biodiversity of the seas and oceans.

A key part of marine science is sustainable natural resource use particularly in relation to fisheries, extraction of fossil fuels and minerals as well as activities like shipping, tourism and other social and human science aspects. The oceans play a key role in climate regulation and biogeochemical cycles which are vital for all life on Earth, as well as providing a range of other ecosystem services.

Key anthropogenic (manmade) threats to the marine environment include;

- Pollution – from industry, agriculture and domestic (sewage) sources resulting in poor water quality, over enrichment with nutrients (eutrophication which can lead to toxic algal blooms), deoxygenation, altered chemistry (pH, salinity), increased suspended sediment loads.

- Ocean warming – leading to altered geographic ranges and abundances of marine and coastal species.
- Ocean acidification – which causes calcium carbonate shells of marine organisms (plankton, shellfish, biogenic reefs) to dissolve.
- Sea level rise – resulting in loss of intertidal habitats (fish nursery grounds, bird breeding and feeding grounds) and natural sea defences for coastal communities.
- Plastic pollution – physical presence of plastics and microplastics in oceans and seas as well as the entanglement of, and ingestion by, marine fauna (e.g. birds, marine mammals, fish, turtles). These also contribute to the presence of microplastics as they degrade.
- Introduction/spread of non-native species, disease and pathogens – these often outcompete and displace native species.
- Loss of species – due to overexploitation of fish and shellfish and direct and indirect take in fisheries (sharks, turtles, marine mammals, birds).
- Loss of coastal habitats due to tourism developments and aquaculture.

10.2 Issues and Trends

Marine environments have historically been overlooked when compared with their terrestrial and freshwater counterparts; in particular their resilience to overexploitation and pollution was severely overestimated. Recognition of the need for marine conversation and protection has notably increased in recent decades. This is reflected in international agreements and initiatives as well as national, regional and international policy and legislation. A global example of this is the United Nation Sustainable Development Goal 14: Conserve and sustainably use the oceans, seas and marine resources for sustainable development.

Coupled with this progress have been vast improvements in research and development (R&D), data collection and analysis and management techniques for these often remote and inaccessible ecosystems. The widespread advocacy of the need for integrated, ecosystem wide approaches to marine conservation management and protection has led to a broadening of the nature and diversity of roles in this sector.

Mumby (2017) outlines some of the key trends in marine science:

> Marine science has made major advances in the past few decades. . .This progress notwithstanding, there are persistent challenges in achieving an understanding of marine processes at appropriate scales and delivering meaningful insights to guide ocean policy and management. . . New technology has allowed marine scientists to make great progress in understanding contemporary biological mechanisms as well as historical processes. . .one of the most significant developments in recent years is the surge in publicly accessible data, particularly from satellites. . .providing unprecedented opportunities to measure the dynamics of coastal ecosystems including seagrass beds, mangroves and coral reefs. . .
>
> Satellites are also revolutionising the management of offshore fisheries. . . Near real-time measures of ocean primary production are allowing fisheries managers to

restrict vessel access dynamically. . .The growth in data availability has been paralleled by advances in computing and collaboration. . .Today, simulated datasets on the hydrodynamic dispersal of marine larvae — so-called 'connectivity data' — have become commonplace. . . These connectivity data have led to diverse applications including the design of marine reserves and strategic targeting of eradication of pest species. . .

Advances in genetics are making a marked impact on our understanding of marine ecosystems. At one extreme, falling laboratory costs have allowed data-intensive sampling practices to become economically feasible. . . At the other extreme, the emergence of environmental DNA (eDNA), whereby traces of organisms in the water column. . . provide a remarkably sensitive way to detect the presence of species, including those that are rare and therefore difficult to sample through conventional means. . .

Biogeochemical tools, such as stable isotopes, continue to provide important insights into the ecology of marine ecosystems. . . Ecologists frequently struggle to measure processes over appropriate time scales, typically because research grants rarely extend beyond five years. Yet, innovations in geological dating techniques are now closing critical gaps on the scale of years to decades.

The rise of mass open source data, innovation in measuring biological processes and collaborative approaches to analysis is bound to create a synergy that will accelerate the pace of discovery in marine biology. . . But scientific progress needs to be accompanied by a nuanced public and political dialog that resists crass scenarios and seeks the best long-term outcome for biodiversity and the vast number of humans that rely on the oceans.

Mumby (2017) also highlights other issues and trends which raise questions around:

- How to manage micro plastics and plastic trash
- The impact of underwater noise on the marine environment – within the field of bioacoustics
- Environmental stress in marine organisms
- Complexity of biogeochemical systems and cycles when undertaking climate change research
- Global policies versus local protection

10.3 Key Organisations and Employers

There are many organisations and employers in the marine sciences sector, which include:

- Government departments such as regulators and statutory bodies, public bodies, laboratories and research bodies
- Not for profit organisations and Non-governmental organisations (NGOs) such as conservation charities

- Commercial companies such as marine energy companies, surveying companies, shipping firms
- Environmental and engineering consultancies
- Universities and academic bodies
- International bodies such as the International Maritime Organisation, an agency of the United Nations

10.4 Careers in the Sector

Globally the demand for marine environmental services is growing. There is an ever-increasing awareness of and need for international and national protection and sustainable management of marine and coastal ecosystems globally. In addition, there is increasing pressure on and drive from industry to make reforms.

Policy and legislation are facilitating this reform (such as the UN Sustainable Development Goal (2015) mentioned earlier).

All of these factors drive demand for marine scientists of all biological, physical, chemical and human disciplines.

Water Encyclopedia (2021) notes:

> The marine sciences offer many employment opportunities. Those interested in pursuing research careers will find opportunities in academia, industry, government, non-profit and nongovernmental organisations, consulting firms and, in some cases—like aquaculture, the growing of marine life for food and medicines—owning their own businesses. . . In an academic setting, most jobs involve conducting research and teaching undergraduate and graduate students and require at least a master's degree and usually a Doctor of Philosophy (Ph.D.) degree.

Social media channels are helpful in terms of finding NGOs that work with species and habitats of interest or are in a specific geographic region. Many are paid voluntary roles that then allow progression; some have internships, and some have paid roles.

10.5 Job Titles in the Sector

Job titles within the sector include:

> Marine Biologist, Marine Scientist, Consultant Marine Ecologist (further special-isms include marine mammals, ornithology, fish, shellfish, marine reptiles), Marine Environmental Consultant, Research Associate, Lecturer, Marine Conservation Manager, Marine Warden, Volunteer, Oceanographer (further specialisms include chemical, physical, geological and biological), Oceanographic Researcher, Ocean Modeller, Marine Physical Processes modeller, Zoological Field Assistant.

10.6   Educational Requirements

- Many roles ask for a degree in a relevant subject such as marine sciences, oceanography or a natural sciences degree (environmental sciences, geography)
- Some roles seek a higher-level degree
- Volunteering, internship or work experience is always valuable, citizen science programmes are offered all over the world and offer great experience as well as networking opportunities.

10.7   Personal Attributes and Skill Sets

Skills and personal attributes may include:

- Knowledge of marine sciences
- Knowledge of legislation, policy, research and practice
- Data analysis, GIS, modelling
- Technical report writing
- Analysis of study documents
- For some roles you may need specific qualifications such as protected species licences
- Communication and result dissemination skills
- Teamwork
- High level IT skills
- Accuracy and research abilities
- Project management
- Leadership and financial management
- Experience of fieldwork and data collection
- Passion and enthusiasm
- Ability to consider work overseas and being flexible to travel for work

Training/Continuing Professional Development (CPD)/ Professional Memberships

There are a number of online introductory and longer-term courses offered by organisations such as the ZSL and National Geographic and even virtual internships such as those offered by GVI which can help you access and progress within this sector alongside more vocational training.

ZSL (Zoological Society of London)
www.zsl.org/united-for-wildlife-free-conservation-courses

National Geographic
www.nationalgeographic.org/projects/exploring-conservation

GVI Virtual Internships
www.gvi.co.uk

Chartered Institute of Ecology and Environmental Management (CIEEM)
www.cieem.net

The Institute of Marine Engineering, Science and Technology (IMarEST)
www.imarest.org/events-courses/training-courses

10.8   Career Paths and Case Studies

◀ **Top Tips**

Rachel Antill, Senior Marine Consultant, APEM Ltd, United Kingdom

1) Develop skills in computer programmes that will be valuable to your employer such as GIS and statistical programmes such as PRIMER, R, SPSS and ensure you understand the mathematics behind them – not just which buttons to press!
2) Get an internship whilst at university as it will improve your chances of getting a job with any firm once you graduate.
3) Try to get work conducting surveys early on in your career. Once you become a desk-based consultant it can be very difficult to get this experience and it is invaluable to have a good understanding of what is involved when undertaking surveys.
4) Apply for jobs using a combination of sector-specific recruitment agencies and applying directly to the firms you wish to work for.
5) Be persistent! In particular, if you don't receive a response to your speculative job application, find another contact within the firm to ensure your application reaches the right person.

Source: Rachel Antill, Senior Marine Consultant, APEM Ltd, UK. © John Wiley & Sons.

Personal Profile

Angela Lowe, Principal Marine Consultant and Director, Medley Marine Limited, United Kingdom

After graduating (and completing the almost mandatory voluntary conservation roles overseas) I wasn't sure how to 'start my career' or where to look for work. I'd been told I wouldn't be able to go straight into consultancy, but it turned out that wasn't the case. My first two positions were in small, independent and highly specialised consultancies; it was very much a baptism of fire – sink or swim. Over the following 14 years I've worked for consultancies of all sizes and worked on large marine and coastal infrastructure projects all over the world.

My specialism is marine ecology, but I have a good working knowledge of all offshore disciplines. My work primarily consists of Environmental Impact Assessments (EIAs) and Lender Compliance work in relation to marine biodiversity and ecosystem services (e.g. International Finance Corporation Performance Standard 6). I really enjoy working in multidisciplinary teams of environmental specialists and engineers. It is important to me to keep up to date with the latest research and development outputs so I can apply them in project contexts. I love being a consultant. It is dynamic, challenging and often extremely pressurised but never dull.

◀ **Top Tips**

1) Find your niche. The environmental sector is large and can be accessed through lots of different channels from internships to PhDs each leading to slightly different, but interlinked paths. Play to your strengths and interests. Look at options to gain relevant work experience not just qualifications.
2) Be passionate. Our line of work can be quite depressing at times, so it is important to do something that you feel is valuable and worthwhile.
3) Keep learning. Be proactive and invest in your continuing professional development. You'll also find that different roles will offer different opportunities to develop and diversify your skillsets. Don't be afraid to move jobs to continue progressing in your career.
4) Network. It's fun, mutually beneficial (we're all on the same team) and ensures you are open to potential opportunities for work or collaboration. Workshops and conferences are particularly valuable for keeping up to date with all the latest developments in your sector.
5) Persevere! I've lost count of the number of days I spent early on in my career speculatively submitting my CV, in person and via email, to companies I wanted to work for. I've since worked for several of them. Environmental work can be very competitive but believe me it can also be very rewarding.
    www.medleymarine.co.uk
    www.linkedin.com/in/angela-lowe-18618322

Source: Angela Lowe, Principal Marine Consultant and Director, Medley Marine Limited, UK. © John Wiley & Sons.

10.9 External Resources

United Kingdom

Marine Biological Association
www.mba.ac.uk
Marine Conservation Society
www.mcsuk.org
Marine Management Organisation
www.gov.uk/government/organisations/marine-management-organisation

IMarEST, The Institute of Marine Engineering, Science and Technology
www.imarest.org

Prospects job profiles
www.prospects.ac.uk/job-profiles/marine-scientist
www.prospects.ac.uk/job-profiles/oceanographer

www.prospects.ac.uk/job-profiles/environmental-education-officer
www.prospects.ac.uk/job-profiles/environmental-consultant
www.prospects.ac.uk/job-profiles/environmental-manager

Europe

EurOcean – The European Centre for information on Marine Science and Technology
www.eurocean.org/np4/home

MareNet 'Marine Research Institutions in Europe'
www.marenet.de/MareNet/europe.html

European Marine Biological Research Centre (EMBRC-ERIC)
www.embrc.eu

European Marine Board
www.marineboard.eu

European Commission 'Marine and Coast – Our Oceans, Seas and Coasts'
www.ec.europa.eu/environment/marine/research/index_en.htm

Asia

MareNet 'Marine Research Institutions in Asia'
www.marenet.de/MareNet/asia.html

Ryan Martinson, The Jamestown Foundation China Brief Volume: 19 Issue: 2 'Assessing the
Future of Chinese Sea Power: Insights from the "Marine Science and Technology Award"'
www.jamestown.org/program/assessing-the-future-of-chinese-sea-power-insights-from-
the-marine-science-and-technology-award

Fuze Ecoteer
www.fuze-ecoteer.com/conservation-jobs-in-asia

Africa

African Conservation
www.africanconservation.org/jobs

MareNet 'Marine Research Institutions in Africa'
www.marenet.de/MareNet/africa.html

UNESCO-IOC. (2009) 'African Oceans and Coasts' Odido M. and Mazzilli S. (Eds). IOC Information Document, 1255, UNESCO Regional Bureau for Science and Technology in Africa
www.unesdoc.unesco.org/ark:/48223/pf0000185095

Paul Kimanzi, ScienceAfrica (2020) 'Advancing Marine Science: Ocean Literacy is Crucial
for Future Generation'
www.scienceafrica.co.ke/advancing-marine-science-ocean-literacy-is-crucial-for-future-
generation

Cochrane KL, Sauer WHH, Aswani S. Science in the service of society: Is marine and coastal science addressing South Africa's needs?
S Afr J Sci. 2019;115(1/2), Art. #4418, 7 pages. https://doi.org/10.17159/sajs.2019/4418

North America

Association for the Sciences of Limnology and Oceanography
www.aslo.org/who-we-are-summary

Marine Conservation Institute
www.marine-conservation.org
www.marine-conservation.org/who-we-are/jobs/career-resources

MarineBio Conservation Society:
Marine Biology Degree Programs in the U.S. (by State)
www.marinebio.org/careers/us-schools

Sea Grant guide to marine careers
www.marinecareers.net

Sea Grant Internship Opportunities
www.seagrant.noaa.gov/Students/Internships

Department of Wildlife and Fisheries Sciences Job Board
www.wfscjobs.tamu.edu/job-board

Ocean Conservation Society
www.oceanconservation.org

Woods Hole Oceanographic Institution
www.whoi.edu

South America

University of Texas at Austin Latin American Network Information Center 'Marine Studies'
www.lanic.utexas.edu/la/region/marine

UN Environment Programme (UNEP) (2018) 'Latin American and Caribbean countries champion marine conservation'
www.unenvironment.org/news-and-stories/story/latin-american-and-caribbean-countries-champion-marine-conservation

The Pew Charitable Trusts (2020) 'Scientist Tackles Plastic Waste Along Latin America's Pacific Shores'
www.pewtrusts.org/en/research-and-analysis/articles/2020/04/09/scientist-tackles-plastic-waste-along-latin-americas-pacific-shores

Latin American Fishery Fellows (LAFF) program
www.sfg.msi.ucsb.edu/share/latin-american-fishery-fellows-program

The Ocean Foundation (2020) 'The Ocean Foundation and University of Havana's Marine Research Center: 21 Years of Science, Discovery, and Friendship' www.oceanfdn.org/the-ocean-foundation-and-university-of-havanas-marine-research-center-21-years-of-science-discovery-and-friendship

Oceania

Australian Institute of Marine Science
www.aims.gov.au

Marine Science Australia
www.ausmarinescience.com

National Marine Science Committee
www.marinescience.net.au

OceanWatch Australia
www.oceanwatch.org.au
www.oceanwatch.org.au/latest-news/marine-science/6-australian-marine-science-trail-blazers-who-are-women

Australian Marine Sciences Association Inc. (AMSA)
www.amsa.asn.au

Australian Marine Parks 'Marine science program'
www.parksaustralia.gov.au/marine/management/programs/marine-science

New Zealand Marine Sciences Society (NZMSS)
www.nzmss.org

Rebecca Jarvis and Tim Young, The Conversation (2019) 'Study identifies nine research priorities to better understand NZ's vast marine area'
www.theconversation.com/study-identifies-nine-research-priorities-to-better-understand-nzs-vast-marine-area-119547

Global

Worldfishing.net – Global Fisheries and Aquaculture Organisations
www.worldfishing.net/directory/categories/organisations

Intergovernmental Oceanographic Commission (2015) 'Transfer of marine technology: knowledge sharing and capacity development for sustainable ocean and coastal management'
www.unesdoc.unesco.org/ark:/48223/pf0000232586

UN – National Governmental Agencies that Deal with Oceans and the Law of the Sea
www.un.org/depts/los/Links/Gov-Agency.htm

UN World Oceans Day
www.unworldoceansday.org
Seven Seas Media – Marine Conservation Jobs
www.sevenseasmedia.org/marine-conservation-jobs

Open Channels – the community hub for sustainable ocean management and conservation
www.openchannels.org

Women in Ocean Science (WOS)
www.womeninoceanscience.com

Wise Oceans
www.wiseoceans.com/jobs

The Society for Conservation Biology (SCB)
www.conbio.org

Earthworks Jobs
www.earthworks-jobs.com/oceanogr.htm

International Maritime Organization
www.imo.org

FOR UPDATED AND ADDITIONAL RESOURCES, INCLUDING EXTRA CHAPTERS, GO TO WWW.ENV.CAREERS

References

Educalingo.com (2020). Definition 'Marine Science'. www.educalingo.com/en/dic-en/marine-science (accessed 14 April 2021).

Environmental Science.org (2021). What Is Oceanography? www.environmentalscience.org/career/oceanographer (accessed 14 April 2021).

Marine Conservation Institute Career Resources (2020). www.marine-conservation.org/career-resources (accessed 14 April 2021).

MarineBio Conservation Society (1998–2021). What is marine biology? www.marinebio.org/creatures/marine-biology (accessed 14 April 2021).

Mumby, P.J. (2017). Trends and frontiers for the science and management of the oceans *Current Biology* Elsevier.

National Geographic (2019). Resource Library – Oceanography. www.nationalgeographic.org/encyclopedia/oceanography (accessed 14 April 2021).

Prospects Job Profile: Marine Biologist (2019). www.prospects.ac.uk/job-profiles/marine-biologist (accessed 14 April 2021).

Prospects Job Profile: Marine Scientist (2019). www.prospects.ac.uk/job-profiles/marine-scientist (accessed 14 April 2021).

Science Daily (2021). Reference terms. www.sciencedaily.com/terms/marine_conservation.htm (accessed 14 April 2021).

UN Sustainable Development Goals (2015). www.unenvironment.org/explore-topics/sustainable-development-goals/why-do-sustainable-development-goals-matter/goal-14 (accessed 14 April 2021).

Water Encyclopedia (2021). Careers in Oceanography. www.waterencyclopedia.com/Bi-Ca/Careers-in-Oceanography.html (accessed 14 April 2021).

11

Protected Area Planning and Management

11.1   Sector Outline

Protected Area Planning, in brief, is the formal processes of planning and implementing protected area policies in a protected area given the values encompassed by the area, the pressures on these values, and taking into account local circumstances, as well as public, Indigenous and stakeholder opinions.

Protected area planning is the first step in effective management of natural, cultural and tourism values at the site-specific level. Planning outcomes (e.g. park management plans and project environmental impact assessments) are what the government or the non-governmental organisation implements when as it protects land and local values by mitigating and preventing negative impacts to these values.

11.2   Issues and Trends

The International Union for Conservation of Nature (IUCN) has been leading the development of protected area planning and management standards and guidelines. The work is ongoing. Many protected area organisations at national and sub-national levels are endeavouring to follow these standards and guidelines. For example, many organisations are taking the first step of measuring the effectiveness of their protected area planning and management frameworks and process.

Important issue – the move towards defining the professionalism within the sector through 'Protected Area Practitioners'. IUCN (2016) 'A Global Register of Competences for Protected Area Practitioners – A comprehensive directory of and user guide to the skills, knowledge and personal qualities required by managers, staff and stewards of protected and other conserved areas'

IUCN (2016):

> If the growing global protected area system is to meet these expectations and to be more than a network of 'paper parks', we need to raise the profile of protected area management as a distinct, formally recognised, respected profession. The ultimate goal of professionalisation is to strengthen individual and organisational performance,

Global Environmental Careers: The Worldwide Green Jobs Resource, First Edition. Justin Taberham.
© 2022 John Wiley & Sons Ltd. Published 2022 by John Wiley & Sons Ltd.

and Foreword thereby the effectiveness of protected areas. 'Professionals' need not only be government staff; they can and do include a wide range of protected and conserved area custodians, including local and indigenous community members, nongovernmental organisations and private owners. Whatever their affiliation, protected area practitioners should be regarded as respected professionals with distinct skills, recognised in the same way as health workers, teachers and engineers. The process of professionalisation includes the adoption of recognised standards of competence and performance, standards that are integrated into qualifications, professional development, career paths and performance assessments, as well as organisational culture and practices. The Global Register of Competences for Protected Area Practitioners provides an essential foundation for this process. In essence, it defines all the possible skills, knowledge and personal qualities required by people working in protected areas around the world. It is an ideal reference and starting point for managers and human resource professionals to plan and manage staffing of protected areas, for educators to identify and meet capacity needs, and for individuals to assess and develop their own skills. Just as importantly, the register demonstrates that ensuring the future of the planet's biodiversity and life support systems is a complex, multi-skilled profession, worthy of respect, recognition and support.

11.3   Key Organisations and Employers

National, regional and municipal governments all have their own protected areas, and many non-governmental organisations such as conservancies and land trusts also manage networks of protected areas. Other NGOs, such as the National Trust in the UK, have considerable protected area networks.

11.4   Careers in the Sector

Within the sector there are many areas into which a career can develop. Often, many people focus on site management or field work, but there are openings in policy and plan development and implementation, general management, environmental education and tourism and recreation.

In terms of organisations involved in the sector, these range from National and State Bodies which manage protected areas (such as Parks Agencies); to resource agencies (especially in forestry and fisheries); NGOs and consultancies.

11.5   Job Titles in the Sector

The IUCN standards and guidelines separate the sector into four personnel types – Skilled Worker, Middle Manager/Technical Specialist, Senior Manager and Executive. It then splits these into competence groups (Planning, Management and Administration; Applied

Protected Area Management; General Personal Competences) then competence categories within each group (which includes Protected Area Policy, Planning and Projects; Biodiversity Conservation; Field/Watercraft and Site Maintenance).

Within the sector there are many roles, including Ranger, Scientific Officer, Tourism Officer, Community Outreach Officer, Park Warden and Site Manager.

11.6 Educational Requirements

Protected Area Planning is a unique field of work for which there are very few academic programs that lead directly to acquiring all the skills used in the job. For example, degrees in Planning are often more geared towards Urban Planning. While many concepts and procedures are common across planning disciplines, there are educational institutions that cover natural resource and conservation planning within disciplines such as environmental science, biology, or geography. As such, be careful to investigate university and college degree courses before choosing the best degree for you. If you are able, Master's degrees and specific post-graduate training courses or certificates are further opportunities to practise self-directed learning and to refine your more generalised degree-level knowledge. While on your educational journey, seize any opportunity to volunteer and/or job-shadow at an organisation you look up to as somewhere you would like to work after your post-secondary education is complete.

11.7 Personal Attributes and Skill Sets

Technical skills and knowledge you will need include the ability to undertake background research, data analysis, committee reporting, project management, teamwork and report writing. Planning procedures can often take a long time and be very complex: patience, empathy and meticulousness are important personal attributes.

Key Duties and knowledge required for Protected Area Planning

Protected area planners and managers work for various levels of government or non-governmental organisations that manage land. Each protected area presents a different set of values and a different land use history, so each project will result in a new experience. As the planner works to achieve the purpose, vision and objectives for that protected area, the tasks they undertake are varied, and include:

- Interpretation and implementation of legislation and policy
- Field trips
- Background and historical research
- Team and committee work
- Outreach and public consultation
- Writing management plans, assessments, e-mails and briefing notes
- Environmental impact assessment

A Protected Area Planner must understand, interpret and implement local, provincial and federal legislation and policies, as well as an understanding of how lands are managed and administered (e.g. an understanding of various forms of land tenure). Protected Area Planners must also learn how non-protected areas and natural resources, such as forests, water and geological resources are managed in the areas surrounding a protected area, for when they need to communicate with other areas of government that manage development of human infrastructure and of natural resources.

In various ways, Protected Area Planners ensure that legislation and policy is applied, that strategic and policy commitments are met, and that social and environmental accountability is considered for the protected areas in their administrative area. To do this, a planner will use their strong analytical and synthesis skills to write management plans and associated site-specific policies that protected area managers will use to help guide programs, development and enforcement. Protected Area Planners also often delineate or amend protected area boundaries. Protected Area Planners also work with protected area managers to undertake environmental impact assessments for development proposals on protected area lands. Unless they are administrative in nature (e.g. correcting errors), these tasks require public consultation. For example, as part of the management planning process, a planner will communicate their in-depth research in a written background document that the public can use to learn about a specific protected area and the management planning process, so that the public may contribute to the process if they wish.

The government must approve before anything is released for public review, so a planner must have a good sense of current issues and the government's priorities of the day to write briefing notes. An understanding of how your government works is a first step, and political acuity is an asset.

As they manage the project from beginning to end, a Protected Area Planner must keep the big-picture goals in mind. The goals are ubiquitous, no matter where in the world you work: the conservation of natural, cultural and tourism values for the benefit of the people and to maintain and enhance ecological integrity. There is a reason why the best places to live are closest to nature – people value nature and wild spaces, and protected areas are becoming more important as human consumption and development expand and as our climate changes. Ecological integrity is hard to replace, once lost. That protected area planning plays an important role in mitigating human impacts on the environment and preserving natural and cultural heritage for future generations makes it a worthwhile career choice.

Important skills and competencies for a Protected Area Planner

Because a Protected Area Planner must liaise with government personnel in other areas of government and must consult public opinion on changes to lands or protected area policies, they must have superior communication skills, be diplomatic and be able to consider various viewpoints.

An understanding of how to go about learning new things is a key competency, and the enjoyment of learning makes the challenge worthwhile. For example, in certain areas, a planner will benefit greatly from learning local Indigenous history, as well as understanding current rights, title and treaties. The Protected Area Planner must ensure that a protected area remains true to its purpose and vision. Each project is different than the last, which ensures a Protected Area Planner is always learning, adapting and gaining new experience. In fact, a love of learning is a common quality of any Protected Area Planner, no matter their educational background.

11.8 Career Paths and Case Studies

Many protected area planners begin with internships or other types of short contracts. The government usually has a standardised recruitment process that any applicant must follow exactly to be included in the competition, and for the most part, an applicant should apply to every position that fits their interests and skillset, no matter the location. Once you are in a government position, you may be able to apply for a wide range of positions closer to your preferred location.

It can be difficult to find volunteer positions in government offices, but there are many conservation not-for-profit organisations (e.g. land trusts and conservancies) that hold similar roles and rely on volunteers. It is always possible to approach people in areas that interest you and ask for advice and guidance. It is always good to know people in your chosen field, so networking is a good idea. Many administrative positions such as Protected Area Planners are filled from people who have started out working 'in the field' in a role other than planning. Seasonal summer student positions in a specific protected are more easily available job opportunities for those looking for their first job experiences in an organisation. Once inside the organisation, job-shadowing opportunities are often available.

Park Planners can work many years at the same position: each park planning project they complete will provide them with professional development and learning opportunities. It is a benefit of the job that, although each project can take months to years, there are always fresh challenges. Less experienced planners rely heavily on the experience and knowledge of more senior planners as they tackle multiple ongoing projects at once. Eventually, they are the ones providing information to those with less experience. In this sense, park planning is a career as opposed to a job. With enough experience, an assistant park planner can apply for more senior park planner positions in the same office or at the head office of their organisation. Because of their knowledge, willingness to learn and versatility, planners can easily move laterally or up into other types of positions (e.g. research-oriented positions, policy and program development, management).

◀ **Top Tips**

Alissa Moenting Edwards, Park Planner at Ontario Ministry of Environment Conservation and Parks, Canada

1) Focus on core communication skills, including writing, speaking and teamwork. Practise and garner feedback as much as possible and any way you can. Planning often requires interacting with others, so understand communication etiquette and make the best impression with your knowledge of good spelling and grammar.
2) Knowing where to start is half the challenge: know how to learn and research any given topic. Keep up with current affairs and political issues of the day, especially with respect to the environment.

3) Understanding some fundamentals of ecology and human impact on the environment (at all scales). Know the statistics and basic trends around these impacts. Be inclusive in your thinking: being conservation-minded is important but understanding and considering a variety of viewpoints helps you understand fundamentally why humans have the impact we do.

4) Governments (and planners who work for them) serve the public: read up on international protected area management trends and how different governments tackle contentious issues and conservation commitments given the many facets of public opinion.

5) Gain an understanding of your local legislation and policies covering protected areas, species-at-risk, conservation and natural resources, which should be available on your government's website. If possible, contact and/or meet a local protected area planner to ask for their knowledge and advice. Ask questions: it can never hurt to show that you are keen and willing to learn.

Alissa Moenting Edwards has a BSc. in Wildlife Biology (minor in Geography) from the University of Guelph and an MSc. in Ecology from Lakehead University. She currently works at the Ontario Parks Northeast Zone Office in Sudbury, Ontario.

Source: Alissa Moenting Edwards, Park Planner at Ontario Ministry of Environment Conservation and Parks, Canada. © John Wiley & Sons.

Organisational Profile

Ontario Parks

Ontario Parks is the managing body for over 600 provincial parks and conservations reserves in the province of Ontario, Canada. Ontario Parks' funding comes from both the government budget and park users, who fund most of the operating costs for those parks that charge fees for day-use and camping. Ontario Parks' head office is in Peterborough, Ontario, and it has five administrative zones: Southeast, Southwest, Algonquin, Northeast and Northwest. Protected Area Planners are employed at both zone and main offices.

Commitments to the Aichi Targets may lead to more park planning positions in committed countries. For example, the Ontario Government has recently voiced support for Canada's commitment (Canada Target 1) to Aichi Target 11 of the Aichi Biodiversity Targets, which commits to protecting at least 17% of terrestrial and inland water, and 10% of coastal and marine biodiverse areas.

◀ **Top Tips**

Audrey Boraski, Hilltown Land Trust, USA

1) Volunteering

When volunteering in your desired field of work you can gain so many benefits personally and professionally. As a volunteer you are not only selflessly helping with an external organisation and serving for someone besides yourself, but you are also making connections and meeting people who could potentially have similar interests to you. Learning new knowledge as well as skills is another advantage of volunteer work. These volunteer experiences and associated skills are transferable to future jobs and can fill your resume. Though volunteering in areas you are interested in starting a career in is helpful, volunteering in other sectors and organisations that fulfil you personally can help make you a well-rounded candidate at job interviews and on applications. This can help because it diversifies your resume and adds other 'soft skills' that are needed for jobs you are applying for.

2) Creating an online presence

With everything turning to the Internet within the last 15–20 years, growing your online identity is equally as important as having a connection and networking in person. Creating a LinkedIn page, your own personal blog or website, and a Twitter profile that is dedicated to an environmental professional person are a few popular and free options to begin this online presence. Adding/following/friending/connecting with individuals who have similar career paths or interests you desire to shadow helps grow your online identity as well as grow connections. Reaching out to people on these platforms is also an encouragement in learning more about their work. Other profiles can also show you easily how they achieved their career goals and the associated paths they took to get there which could save you time in your own.

3) Joining organisations

Joining organisations as a member either locally, nationally, or internationally can be a powerful tool in your career journey. Some small organisations can include a local land trust or cities conservation commission while larger ones can include The Wildlife Society or WWF. Not only does joining these as a member tell you more about organisations that you are interested in working for, but it also keeps you up to date on the latest news in those fields. Joining organisations as a member, whether it is free or costs an annual fee, can also lead to discounts on conference prices as well as help you network and grow connections in a different way. Being a member of an organisation or society also allows you to add something else to your resume and LinkedIn profile. This tip can have similar benefits to my first tip, volunteering, depending on the type of membership you decide to participate in (e.g. Active board committee member vs. student/national member).

Keeping up with the news in your field of interest in the environmental sector is critical. Subscribing to newsletters of organisations and agencies that you are interested in is a quick and easy way to stay on top of the new findings and work being done around the world. Making a point to read these newsletters and subscriptions every time they are sent to you can be helpful as well as saving them in a folder on your email. Listening to podcasts, reading books, joining book clubs and watching documentaries that are relevant to the environment are also all good ways to take in information on news.

4) Your resume

Another great tip I was given once, and now preach, was to always have an up-to-date resume on you. This can be very important because some job postings pop up quickly and only consider the earliest applicants, so getting your application submitted as quickly as possible is essential. Having a running resume of everything you have done and one or two more specific to environmental areas you are interested in is extremely helpful for quick applying. I would recommend updating your resume/CV every time you have new content to add as well as a deeper organisation every three months or every six months to keep it perfected. If you were in college/university and have a Career and Academic Advising Center or Writing Center this can be a free service. If this is not the case, having an advisor or supervisor or a friend or family member who is willing is just as good. Either having these resumes on a USB with you always or saved online somewhere to access can be a helpful way to have them ready to go.

5) Patience and openness

Practicing patience and openness is extremely important in the environmental sector. Looking at your career journey with a wider lens is crucial, especially for beginner job seekers. It is essential to frequently visualize and reflect on how far you have come because it can feel like a slow journey to your perfect career. Reflecting on your journey will help you celebrate the small victories that you might pass over when they occur, which is important for your mental health. Patience is very important because if you try to work too much you will overdo it and burn out, potentially breaking ties with this field of work. With this all being said, jobs that serve the environment are some of the most rewarding out there so stick to what makes you happy every day!

Personal Profile

Growing up surrounded by dense forests and varying geographical features fostered my love for nature and sparked my curiosity about the critters all around. Wanting to learn more about the living world, I carried my curiosity through adulthood by earning a Bachelor of Science in Biology and I am currently finishing a Master of Science in Conservation Biology. Simultaneously with school, I am currently working as a Land Stewardship Coordinator at Hilltown Land Trust. Here, I work to preserve land in conservation, including the sensitive environments and species within, to create connected habitat for all life to thrive and enjoy. My previous work has taken me from the insect-packed interior of the Great Dismal Swamp to the sun-filled rolling dunes of Cape Cod. I have worked multiple seasons in the Conservation and Ecology sector in roles ranging from environmental researcher to wildlife technician to biology and conservation intern for many different organisations over the past six years.

Audrey Boraski is a Masters student studying Conservation Biology and experienced field technician Currently serving as an AmeriCorps member as a Land Stewardship Coordinator.

Source: Audrey Boraski, Hilltown Land Trust, USA. © John Wiley & Sons.

Company Profile

Hilltown Land Trust, Ashfield Massachusetts, USA

Hilltown Land Trust works to conserve ecologically important wildlands, economically and culturally important working lands, and the scenic beauty and rural character of the hilltowns of Massachusetts, United States.John Wiley & Sons

Our service area includes 13 rural towns: Ashfield, Chester, Chesterfield, Conway, Cummington, Goshen, Huntington, Middlefield, Plainfield, Westhampton, Williamsburg, Windsor and Worthington.

Since our founding in 1986, we have protected nearly 5000 acres of land including 9 properties that we own and 34 Conservation Restrictions (CRs) which remain in private ownership.

www.hilltownlandtrust.org/.

Source: Hilltown Land Trust, Ashfield Massachusetts, USA. Retrieve from www.hilltownlandtrust.org/.

Personal Profile

Mikael Trewick, Land Management Team Officer, Sheffield and Rotherham Wildlife Trust, UK

I am Land Management Team Officer for the Sheffield and Rotherham Wildlife Trust. I also work as a Site Officer for an environmental contractor in my spare time. My Key responsibilities are as follows: – Undertaking practical land management on the Sheffield And Rotherham Wildlife Trusts portfolio of nature reserves. – Policy making (Pesticides, COSHH, Environmental, Machinery) – Health and Safety Documentation – Volunteer recruitment and development. – Leading teams on task and events. – Budgeting – Machinery and hand tool maintenance (This includes Vehicles, Chainsaws, Generators).

◀ **Top Tips**

1) Volunteer- I have over 3000 hours of volunteering across the UK, with the National Trust, Wildlife Trusts, Fix the Fells and so on. It is essential to meet new people, go on taster days, show your interest in surveying or learning new skills.
2) Keep your CV updated. List training courses, meetings, one off training sessions. It shows the breadth of your interests outside of the practical work.
3) Attend workshops provided by organisations such as the Bumblebee Conservation Trust, Moors For The Future, Sheffield and Rotherham Wildlife Trust. Consider signing up for the Countryside Management Association.
4) Get licences and certificates. The big ones are good, but the more obscure can set you aside in a recruitment process. Dry Stone walling, Newt Surveying, ATV driving, brush cutter, Adult Safeguarding, Outdoor first aid. . . It's one less thing for your new employer to worry about.
5) Don't be afraid to apply for jobs and get knocked back. The feedback is vital to your continued development. It'll seem tough, working for free and not getting far initially, but being available and willing to learn sets you aside, and all the hard work pays off.

Source: Mikael Trewick, Land Management Team Officer, Sheffield and Rotherham Wildlife Trust, UK. © John Wiley & Sons.

Personal Profile

Beth Môrafon, Founder and Director, VisitMôr

I've designed and managed over 40 large-scale visitor experiences, bringing the joy of the natural world to international audiences. I've devoted 20 years' practice to visitor centre planning with the Wildfowl & Wetlands Trust and its subsidiary, WWT Consulting. And I've worked with clients from the RSPB to the Wildlife Trusts, along with councils and NGOs across Europe, Africa and Asia. Many UK projects, including an Interpretation Action Plan winning £1.3 million, have been National Heritage Lottery Funded.

11.9   External Resources

UK

National Parks 'National Parks are protected areas'
www.nationalparks.uk/students/whatisanationalpark/nationalparksareprotectedareas

JNCC 'UK Protected Areas'
www.jncc.gov.uk/our-work/uk-protected-areas

IUCN National Committee United Kingdom (2012) 'Putting Nature on the Map'
www.portals.iucn.org/library/sites/library/files/documents/2012-102.pdf

Europe

European Environment Agency (2020) 'An introduction to Europe's Protected Areas'
www.eea.europa.eu/themes/biodiversity/europe-protected-areas

EUROPARC Federation
www.europarc.org

European Environment Agency (2012) 'Protected areas in Europe – an overview'
www.scales-project.net/NPDOCS/EEA%2005-2012%20Protected%20areas%20in%20Europe-an%20overview.pdf

Asia

Juffe-Bignoli, D., Bhatt, S., Park, S. et al. (2014). Asia Protected Planet 2014. UNEP-WCMC: Cambridge, UK.
https://www.unep-wcmc.org/resources-and-data/protected-planet-report-2014

IUCN (International Union for Conservation of Nature) and MOEJ (Ministry of the Environment, Japan) (2014) The First Asia Parks Congress: Report on the Proceedings 13th–17th November 2013, Sendai, Japan
www.asiaprotectedareaspartnership.org/images/apap/apc-docs/apc-proceedings-final.pdf

Rainforest Trust (2019) 'Conservation in Asia-Pacific: Year in Review'
www.rainforesttrust.org/conservation-in-asia-pacific-year-in-review

Abdul Rahoof, Earth.org (2019) 'Protected Areas: the Past, Present, and Future of Conservation'
www.earth.org/protected-areas-the-past-present-and-future-of-conservation

Africa

African Parks 2018 Annual Report 'Unlocking The Value Of Protected Areas'
www.africanparks.org/unlocking-value-protected-areas

African Wildlife Foundation 'Protected Area Expansion'
www.awf.org/land-protection/protected-area-expansion

Wildlife Angel (2018) 'What Model for Wildlife Conservation in Africa: private reserve, conservancy, national park?'
www.wilang.org/en/wildlife-conservation-management-in-africa

European Commission Joint Research Centre (2007) 'The Assessment of African Protected Areas – A characterisation of biodiversity value, ecosystems and threats, to inform the effective allocation of conservation funding'
A.J. Hartley, A. Nelson, P. Mayaux and J-M. Grégoire
www.dopa.jrc.ec.europa.eu/sites/default/files/AssessmentOfAfricanProtectedAreas_EUR22780.pdf

North America

USGS Protected Areas
www.usgs.gov/core-science-systems/science-analytics-and-synthesis/gap/science/protected-areas

IUCN World Commission on Protected Areas North America
www.iucn.org/commissions/world-commission-protected-areas/regions/wcpa-north-america

Government of Canada – Protected Areas
www.canada.ca/en/services/environment/conservation/protected-areas.html

Ontario Parks
www.ontarioparks.com
www.ontarioparks.com/parksblog/5-questions-with-a-park-planner

Ontario – Provincial parks and conservation reserves planning
www.ontario.ca/environment-and-energy/provincial-parks-and-conservation-reserves-planning

Pathway to Canada Target 1
www.conservation2020canada.ca/home

Canadian Parks and Wilderness Society
www.cpaws.org/our-work/parks-protected-areas

South America

Fundación Natura, IUCN Colombian Committee and Parques Nacionales Naturales Colombia (2009) 'Protected Areas and Development in Latin America' Guerrero, E. & S. Sguerra (Editors)
www.portals.iucn.org/library/sites/library/files/documents/2009-046.pdf

Convention on Biological Diversity 'Latin American Alliance To Strengthen Protected Areas' REDPARQUES, Pronatura México (2018). Progreso de cumplimiento de la 11 de Aichi en los países de la Redparques: resultados y

perspectivas al 2020. CDB, Pro-yecto IAPA, Unión Europea, WWF, FAO, UICN, ONU Medio Ambiente. Bogotá, Colombia. 46 p
www.cbd.int/pa/presentations/latin-american-alliance-to-strengthen-protected-areas.pdf

K.D. Thelen and N.G.C. van der Werf, FAO 'Wildlife and rural development in Latin America'
www.fao.org/3/v6200t/v6200T0e.htm

Oceania

Australian Government Department of Agriculture, Water and the Environment 'Australia's protected areas'
www.environment.gov.au/land/nrs/about-nrs/australias-protected-areas

Ben Boer and Stefan Gruber, IUCN (2010) 'Legal Framework for Protected Areas: Australia'
www.iucn.org/downloads/australia_1.pdf

Pew Charitable Trusts (2018) 'Australia to Fund Five New Indigenous Protected Areas'
www.pewtrusts.org/en/research-and-analysis/articles/2018/07/16/australia-to-fund-five-new-indigenous-protected-areas

Department of Conservation 'Parks & recreation'
www.doc.govt.nz/parks-and-recreation

Global

IUCN Guidelines for Management Planning of Protected
www.portals.iucn.org/library/efiles/documents/PAG-010.pdf
Best Practice Guidelines
www.iucn.org/theme/protected-areas/publications/best-practice-guidelines
Convention on Biological Diversity Aichi Biodiversity Targets
www.cbd.int/sp/targets
The United Nations Environment Programme World Conservation Monitoring Centre (UNEP-WCMC) (2014) '2014 United Nations List of Protected Areas'
https://www.unep-wcmc.org/system/dataset_file_fields/files/000/000/263/original/2014_UN_List_of_Protected_Areas_EN_web.PDF?1415613322

Reference

IUCN (2016). A global register of competences for protected area practitioners – a comprehensive directory of and user guide to the skills, knowledge and personal qualities required by managers, staff and stewards of protected and other conserved areas. https://www.iucn.org/content/a-global-register-competences-protected-area-practitioners(accessed 14 April 2021).

12

Waste and Resource Management and Contaminated Land

12.1   Sector Outline

UNEP, in collaboration with the International Solid Waste Association (ISWA), noted in the UNEP/ISWA (2015) Global Waste Management Outlook:

> Inadequate waste management has become a major public health, economic and environmental problem, with 7-10 billion tonnes of urban waste produced each year and 3 billion people worldwide lacking access to controlled waste disposal facilities. . . Waste management is a basic human need and can also be regarded as a 'basic human right'. Ensuring proper sanitation and solid waste management sits alongside the provision of potable water, shelter, food, energy, transport and communications as essential to society and to the economy as a whole. . . Low- and middle-income countries still face major challenges in ensuring universal access to waste collection services, eliminating uncontrolled disposal and burning and moving towards environmentally sound management for all waste. Achieving this challenge is made even more difficult by forecasts that major cities in the lowest income countries are likely to double in population over the next 20 or so years, which is also likely to increase the local political priority given to waste issues. Low- and middle-income countries need to devise and implement innovative and effective policies and practices to promote waste prevention and stem the relentless increase in waste per capita as economies develop.

The Waste Atlas Partnership (2014) is a crowdsourcing free access map that visualises municipal solid waste management data across the world for comparison and benchmarking purposes. In its report 'The world's 50 biggest active dumpsites' it is noted:

> More than half of the world's population is using dumpsites for waste disposal. . . Most of dumpsites are located in Africa, Latin America, the Caribbean and Northern Asian countries, namely in areas where more than two thirds of the Earth's population lives. The sites are historically and physically different. They also differ in size, receive various amounts of waste and host different numbers of

Global Environmental Careers: The Worldwide Green Jobs Resource, First Edition. Justin Taberham.
© 2022 John Wiley & Sons Ltd. Published 2022 by John Wiley & Sons Ltd.

people either working at the dumps or living in the surroundings. However, these 50 sites all have at least one thing in common: the serious threat they pose to human health and the environment. They cannot be considered simply as a local problem. . .Eliminating all such dumpsites around the world must be a priority for the global community.

Waste Management and Contaminated Land are interwoven issues. CIWEM (2020) notes:

> Contaminated land is a global issue and many countries have a legacy of contaminated land resulting from historical industrial activity, waste management and from the previous regeneration of land. Contamination is still occurring across the world, due to inadequate or inappropriate operational controls and regulation in industry and in the management of waste. The presence of contamination can represent an unacceptable risk to human health or the environment.

Globally, there is a mosaic of differing practices and achievements along the line of:

> No formal waste management and dumping of waste – Formal waste management regulations and practices – Migration towards resource management and waste minimisation

Allied Analytics LLP noted in its 2019 report 'Waste Management Market by Waste Type and Service: Global Opportunity Analysis and Industry Forecast, 2018–2025':

> The global waste management market size is expected to reach $530.0 billion by 2025 from $330.6 billion in 2017. . .Waste management is the collection, transportation, and disposable of garbage, sewage, and other waste products. It involves treatment of solid waste and disposal of products and substances in a safe and efficient manner. The growth of the global waste management market is driven by increase in adoption of proactive government measures to reduce illegal dumping. In addition, surge in population and increased globalisation have led to rise in the overall waste volume, worldwide. The urban population produced about 1.3 billion tons of municipal solid waste (MSW) in 2012, which is expected to grow to 2.2 billion tonnes by 2025. Moreover, increase in industrialisation in the emerging economies, such as India, China, and Taiwan, has led to the development of chemical, oil & gas, automobile, and medical industries, which generate enormous amount of waste and cause pollution. These factors are expected to significantly contribute towards the growth of the global market. However, high cost of procuring and operating waste management solutions is expected to hamper the market growth. Conversely, increase in awareness among public and government agencies about these solutions and upsurge in need to develop waste-to-energy solutions are expected to provide lucrative growth opportunities for market players during the forecast period.

12.2   Issues and Trends

There have been significant changes to waste management approaches in the past few decades. There has been a shift of approach to treating waste as a 'resource' rather than as just a waste material.

As noted by UNEP (2015):

> The goal is to move the fundamental thinking away from 'waste disposal' to 'waste management' and from 'waste' to 'resources' – hence the updated terminology 'waste and resource management' and 'resource management', as part of the 'circular economy'.
>
> In fact, the system of landfill mining to recover valuable materials from old electrical equipment is being considered in many areas.

However, the management of waste material globally is a major concern, even if there is a change in focus in the sector. Many regions do not have formalised waste management systems and processes in place. Too much waste management still depends on 'out of sight out of mind' approaches. The global trade in waste material and processing is a good example of this attitude.

The changes within the global sector present opportunities as well as challenges. The Waste Recruit (2020) blog highlights in its article 'Innovations shaping waste and resource management':

> Working in waste management is both challenging and interesting. While many people and organisations may subscribe to the idea of a circular economy, those working in the industry know that the practical steps to make it happen are not that simplistic. Not only do business strategies need to change but with every change comes another practical implementation challenge. It takes continual innovative thinking to overcome these. Fortunately, there seems to be no shortage of innovative minds which is great news for the industry. . .
>
> Typically sorting facilities use infrared imaging to sort different types of plastic for recycling and because black absorbs infrared light the sensors don't pick up [black plastic] packaging and it gets discarded as waste. Unilever. . .set out to solve this problem together with several industry partners. The solution was to change the pigment in the black dye. . . This not only makes the plastics easier to detect, they can also be easily extracted in the recycling process.
>
> Lithium-ion batteries are a headache for the recycling industry often being the cause of fires. Additionally, extracting and recycling the battery materials is not easy. In a pilot project, American Manganese confirmed that independent lab test showed high extraction rates of cathode materials using a new and recently patented technology.
>
> One of the biggest challenges with recycling is the problem of contamination. IBM researchers set out to solve this problem. . .Plastic is ground up and then cooked at temperatures above 200 degrees centigrade. . .producing an end result that is a white powder. The powder can readily be made into new plastic containers. . .
>
> It is exciting to see these innovations and it makes me wonder: "What's next?" With growth happening in all sectors of the industry, there's no doubt that we will continue to see creative solutions to industry challenges.

12.3   Key Organisations and Employers

Within the sector, key organisations include national, local and municipal government as well as consultancies and waste management companies. There are a growing number of roles in regulatory processes, planning and policy making within the sector.

The publication Pulp & Paper Technology (2021) notes the top waste management companies globally as:

Advanced Disposal Services

Advanced Disposal, based in Ponte Vedra, Florida, USA provides non-hazardous solid waste collection, transfer, recycling and disposal services in 16 US states and the Bahamas. There is a pending merger as of 2020 with the company Waste Management Inc.

Biffa Group

A UK-based company with operations at over 195 sites spanning the length and breadth of the country, we service over 2.2 million households and collect 4.3 million bins per week just within our municipal division. From our conception in 1912, our business has developed far beyond waste collection and into recycling, treatment and energy generation services.

Clean Harbors, Inc.

Founded in 1980, Clean Harbors stands to be North America's leading provider of environmental, energy, and industrial services. The company serves a diverse customer base, including a majority of the Fortune 500 companies across the chemical, energy, manufacturing, and additional markets, as well as numerous government agencies. These customers rely on Clean Harbors to deliver a broad range of services such as end-to-end hazardous waste management, emergency spill response, industrial cleaning and maintenance, and recycling services. Through its Safety-Kleen subsidiary, Clean Harbors is also North America's largest re-refiner and recycler of used oil and a leading provider of parts washers and environmental services to commercial, industrial, and automotive customers. Founded and based in Massachusetts, Clean Harbors operates throughout the United States, Canada, Mexico, and Puerto Rico.

Covanta Holding Corporation

Covanta is a world leader in providing sustainable waste and energy solutions. The Company's Energy-from-Waste (EfW) facilities provide communities and businesses around the world with environmentally-sound solid waste disposal by using waste to generate clean, renewable energy.

Hitachi Zosen Corporation

Design, construction and manufacture of Energy-from-Waste plants, desalination plants, water and sewage treatment plants, marine diesel engines, press machines, process equipment, precision machinery, bridges, hydraulic gates, shield tunnelling machines, and equipment for use in disaster prevention/mitigation

---

Remondis AG & Co. Kg

REMONDIS is one of the world's largest recycling, service and water companies. With over 30 000 employees and around 900 business locations on 4 continents, the group serves more than 30 million people and many thousands of companies. The REMONDIS Group operates in many fields of business: it recovers raw materials from waste, develops innovative recycled products, offers alternative fuels and plays an important role in the water management sector supplying water and treating wastewater. In addition, REMONDIS removes pollutants from residual and hazardous wastes – which are unable to be recycled with today's technology – and disposes of them using eco-friendly methods.

Suez Environment S.A.

With 90 000 people on the five continents, SUEZ is a world leader in smart and sustainable resource management. We provide water and waste management solutions that enable cities and industries optimize their resource management and strengthen their environmental and economic performances, in line with regulatory standards. With the full potential of digital technologies and innovative solutions, the Group recovers 45 million tons of waste a year, produces 4.4 million tons of secondary raw materials and 7.7 TWh of local renewable energy. It also secures water resources, delivering sanitation services to 66 million people and reusing 1.1 million m^3 of wastewater.

Veolia Environment S.A.

Veolia group is the global leader in optimised resource management. With nearly 178 780 employees worldwide, the Group designs and provides water, waste and energy management solutions that contribute to the sustainable development of communities and industries. Veolia helps to develop access to resources, preserve available resources, and to replenish them. In 2019, the Veolia group supplied 98 million people with drinking water and 67 million people with wastewater service, produced nearly 45 million megawatt hours of energy and converted 50 million metric tons of waste.

Waste Management Inc.

Houston, USA-based Waste Management, which provides waste management environmental services, owned or operated 247 solid waste landfills as of December 2019. We're the leading provider of comprehensive waste management in North America, offering services that range from collection and disposal to recycling and renewable energy generation.

Republic Services, Inc.

We are an industry leader in recycling and non-hazardous solid waste in the U.S., and our vision is to be America's preferred partner. Our operations focus on providing effective solutions to make responsible waste disposal effortless for our customers.

ISWA – The International Solid Waste Association

ISWA has a mission 'To Promote and Develop Sustainable and Professional Waste Management Worldwide'. According to the ISWA website:

ISWA achieves its mission through:

- Promoting resource efficiency through sustainable production and consumption
- Support to developing and emerging economies
- Advancement of waste management through education and training
- Promoting appropriate and best available technologies and practices
- Professionalism through its programme on professional qualifications.

12.4 Careers in the Sector

This is a sector which is transitioning from a waste-based 'out of sight' approach to a waste minimisation and resource management approach. Consequently, careers within the sector are changing and developing.

12.5 Job Titles in the Sector

Lists of job titles in the different parts of the sector

Waste Management Officer, Waste Technician, Waste Consultant, Operations Manager – Resources/Waste Management, Environmental and Waste Officer, Waste Planner, Recycling Manager, Contaminated Land Officer, Site Manager.

12.6 Educational Requirements

The sector has many roles which have varying educational requirements, from manual roles through to high level scientific careers with require PhD level education. Some roles within the sector ask for specialist qualifications, certifications and training.

The Prospects (UK) website (2020) notes:

> A degree in waste management or similar is often preferred. Other relevant subjects include:
>
> - biological or biochemical sciences
> - chemical and physical sciences
> - civil, structural or mechanical engineering
> - earth sciences
> - environmental science
> - geography and/or geology.
> - Entry is possible with an HND/Diploma in a waste management, environmental protection or a subject related to environmental management.

There is no standard route into this role, although most new entrants are graduates.

Entry requirements have changed in recent years, due to specific waste management courses and an increase in environmental qualifications. While direct entry is common, some people move into this profession after working in the construction, haulage or quarrying industries or by specialising from a wider environmental role within a large organisation.

A pre-entry postgraduate qualification, for example an MSc or PgDip in waste management or environmental engineering, is another route into the sector.

12.7   Personal Attributes and Skill Sets

With the rapid change from a market based on waste disposal to one based on resource management, changing skills needs are a crucial issue.

The 2012 report 'Sustainable Skills: The Future of the Waste Management Industry' by the All Party Parliamentary Sustainable Resource Group, United Kingdom highlighted key skills issues in the sector through a series of expert essays:

> the waste world is facing a new challenge and is in need of one key thing to continue prospering: skills. . . Civil and mechanical engineers are the most common among the professionally qualified staff, employed as landfill designers and site managers. This is supplemented by vocationally trained supervisory and operational staff.
>
> Chemists and environmental scientists complete the current skills landscape, undertaking support functions such as site monitoring. The sector has also had non-landfill operations (notably incineration, composting and hazardous waste management) employing professional chemists and mechanical engineers. . .This is changing.
>
> The new resource-oriented direction of the sector has transformed its skill requirements and altered the dynamics of supply and demand. The deployment of processing technologies has led to a demand for operational staff with a technical, process-related background. With the sector increasingly reliant on the sale of products (recyclates or recovered power) made out of this processed material, there is a growing demand for staff with a background in procurement, sales, and commodity trading.
>
> Skill requirements for managers and supervisors have also changed, with greater emphasis on customer management, product quality, budgetary control, and financial modelling. . . the available skills base is unlikely to keep pace with the expected speed of transformation, and that the sector has to respond by planning for the impending skills gap. . .
>
> Individual companies are addressing this challenge by mapping out a timeline by which specific skills (by type and number of personnel) need to be in place, and then formalising appropriate recruitment and personal development goals and programmes. Companies are collaborating with educational institutions to develop bespoke technical and managerial training courses, as well as to establish an intake of fresh graduates and apprentices to be given in-house training.

Training

There are many training providers in the sector, including those below.

Environmental Expert
www.environmental-expert.com/waste-recycling/waste-management/training

Swana
www.swanaontario.org/training

COMS
www.conference-service.com/conferences/waste-management.html

Learning Cloud
www.learningcloud.com.au/courses/715/environmental-waste-management

IWMSA
www.iwmsa.co.za/training-courses

WAMITAB
www.wamitab.org.uk

Singapore Government Skills Future SG
www.ssg.gov.sg/wsq/Industry-and-Occupational-Skills/Waste-Management-and-Recycling-Industry.html

ISWA
www.iswa.org/events-courses/calendar/

The In-House Training Company
www.in-house-training.com/subject-areas/health-safety-and-environmental/waste-management

Aqua Enviro
www.conferences.aquaenviro.co.uk/training

CIWM
www.ciwm.co.uk/ciwm/training/waste-management-training.aspx

City and Guilds
www.cityandguilds.com/qualifications-and-apprenticeships/built-environment-services/environmental-services/0746-sustainable-waste-management

Waste Training Services
www.waste-training.com

Bywater
www.bywater.co.uk/course/waste-management-training-course

12.8   Career Paths and Case Studies

◀ **Top Tips**

Sarahjane Widdowson, Director, INTELISOS, United Kingdom

Tell anyone that you work in waste and recycling and you can guarantee that they'll have something to say on the subject. Waste is a subject that touches us all. In the United Kingdom, it's a universal service which we duly take part in every week but there's so much more to it than your doorstep waste and recycling collection service. It's an industry that recruits from all disciplines – not just those with an environmental background. It's both a social and scientific discipline.

A career in waste management can range from advising the public on avoiding plastic consumption through to procuring a multi-million-pound waste treatment facility. You could be working for a local authority, private sector contractor, government agency, charity or start-up. It's an incredibly diverse industry.

1) Develop your networking skills
   - Like many sectors developing your career can be helped by getting to know the people already working within the sector. One of the best ways to do this is to join and get involved in the Chartered Institution of Wastes Management. Attend events, introduce yourself to people, join groups on LinkedIn and identify who the influences are sector and follow them on social media. Better still – develop your own voice.
2) Consider practical experience
   - Much of waste management is a 'hands on' discipline and getting practical experience can help you to understand how services work and importantly what the public think and how they behave. Entry level jobs can include recycling advisory, survey work, waste composition analysis and data analysis.
3) Keep up to date
   - The waste sector is a fast-paced environment with a moving policy landscape which makes it both an exciting but challenging area to work in. Keeping up to date is essential so it's important to regularly review trade press sites as well as the national press to understand the current mood.
4) Be creative
   - The sector is ripe for innovation. If you can apply fresh thinking, technology, or behaviour change to improve the environment and protect health there are plenty of opportunities for managing our resources more effectively. The rise of the circular economy is challenging the sector to think differently and will influence how we consume products and materials in the future and therefore how we reuse, recycle and dispose of them.

5) Think globally
  - Waste arises wherever humans have made their mark. Although local regulations and approaches may differ the principles of reducing harm to health and the environment remain the same, and an understanding of these fundamentals will enable anyone working in the sector to apply their skills around the world. Around 1 in 3 people globally don't have access to decent waste management – there are lots of opportunities to do some good.

Sarahjane Widdowson BSc (Hons), MSc, MCIWM, FRSA is a trustee for the charity WasteAid UK.

Source: Sarahjane Widdowson, Director, INTELISOS, UK. © John Wiley & Sons.

◀ **Top Tips**

Mike Tregent, Advisor Waste Strategy, Environment Agency (Resource Magazine Hot 100 Ranked in the Waste and Resources Industry 2018), United Kingdom

1) Learn all the aspects of the sector, plant, commercial, regulatory, etc. You will gain a deeper understanding of other people's perspective!
2) Never stop networking, the sector is like a big family and when people get to know you, you can learn a lot more and be able to influence!
3) Never stop learning, keep up to date with the latest news and events, it's a constantly changing and dynamic sector!
4) Don't be put off by differences, there are always many different opinions out there and often there is more than one answer! Be informed and join the debate!
5) You'll need to be thick skinned, if you want to do the right thing in terms of sustainability and resource efficiency, you will have to be prepared to explain your position clearly and provide evidence!

Source: Mike Tregent, Advisor Waste Strategy, Environment Agency (Resource Magazine Hot 100 Ranked in the Waste and Resources Industry 2018), UK. © John Wiley & Sons.

◀ **Top Tips**

Stephen Wise, Waste Sector Director and Waste Technical Leader at Wood Group, United Kingdom

1) Be flexible
2) Keep expanding your knowledge
3) Develop a good personal network
4) Speak to everyone equally – they all have insights!
5) Be prepared to get your hands dirty!

Source: Stephen Wise, Waste Sector Director and Waste Technical Leader at Wood Group, UK. © John Wiley & Sons.

12.9 External Resources

United Kingdom

Biffa
www.biffa.co.uk/careers

CIWM Apprenticeships
www.ciwm.co.uk/ciwm/apprenticeships/apprenticeships.aspx

Waste Jobs
www.wastejobsuk.com

The Scottish Environment Protection Agency (SEPA):
www.sepa.org.uk/environment

Natural Resources Wales:
www.naturalresources.wales/?lang=en

The Northern Ireland Environment Agency (NIEA):
www.doeni.gov.uk

The Environment Agency in England:
www.gov.uk/government/organisations/environment-agency/about/recruitment

Europe

European Commission – Waste
www.ec.europa.eu/environment/waste/htm

European Environment Agency – Waste Management
www.eea.europa.eu/themes/waste/waste-management

European Parliament News (2018) 'Waste management in the EU: infographic with facts and figures'
www.europarl.europa.eu/news/en/headlines/society/20180328STO00751/eu-waste-management-infographic-with-facts-and-figures

Asia

UNEP (2017) 'Waste Management in ASEAN Countries: Summary Report'
www.unenvironment.org/resources/report/waste-management-asean-countries-summary-report

UNEP, IAT and ISWA 'Asia Waste Management Outlook (AWMO)'
www.wedocs.unep.org/bitstream/handle/20.500.11822/27293/AsiaWMO_Sum.pdf?sequence=1

Hannah Ellis-Petersen, The Guardian UK (2019) 'Treated like trash: south-east Asia vows to return mountains of rubbish from west'
www.theguardian.com/environment/2019/may/28/treated-like-trash-south-east-asia-vows-to-return-mountains-of-rubbish-from-west

Andrew McIntyre, Asian Development Blog (2017) 'Waste management in Asia: 1 goal, 5 cities, 5 lessons'
www.blogs.adb.org/blog/waste-management-asia-1-goal-5-cities-5-lessons

Africa

UNEP (2018) 'Africa Waste Management Outlook'
www.unenvironment.org/ietc/resources/publication/africa-waste-management-outlook

UNEP (2018) 'In pictures: How Southern Africa manages its waste'
www.unenvironment.org/news-and-stories/story/pictures-how-southern-africa-manages-its-waste

UNEP (2020) 'A future in recycling: from street waste collector to entrepreneur'
www.unenvironment.org/news-and-stories/story/future-recycling-street-waste-collector-entrepreneur

North America

James Hamilton, US Bureau of Labor Statistics, Careers in Environmental Remediation, September 2012
US Bureau of Labor Statistics
www.bls.gov/green/environmental_remediation/remediation.htm

Advanced Disposal
www.jobs.advanceddisposal.com

Clean Harbors
www.careers.cleanharbors.com

Republic Services
www.republicservices.jobs

Waste Management
www.careers.wm.com/us/en

South America

UNEP (2018) 'Waste Management Outlook for Latin America and the Caribbean'
www.unenvironment.org/resources/report/waste-management-outlook-latin-america-and-caribbean

UNEP Policy Brief 'Waste management as an essential service in Latin America and the Caribbean' www.wedocs.unep.org/bitstream/handle/20.500.11822/32615/COVID19_WASTE_LAC.pdf?sequence=1&isAllowed=y

Public Services International (2017) 'Municipal Solid Waste Management Services in Latin America'
www.world-psi.org/sites/default/files/documents/research/web_en_lrgm_waste_report_ia_20174.pdf

Oceania

Australian Government – National Waste Policy
www.environment.gov.au/protection/waste-resource-recovery/national-waste-policy

Waste Management and Resource Recovery Association of Australia (WMRR)
www.wmrr.asn.au

CSIRO (2020) 'Circular Economy and Waste Management'
www.csiro.au/en/Research/Environment/Circular-Economy

Australian Bureau of Statistics (2019) 'Waste Account, Australia, Experimental Estimates'
www.abs.gov.au/statistics/environment/environmental-management/waste-account-australia-experimental-estimates/latest-release

Global

UN Jobs
www.unjobs.org/themes/waste-management

ISWA
www.iswa.org/iswa/organisation/about-iswa/

Waste Recruit
www.wasterecruit.com

Remondis
www.remondis.com/en/careers

Suez
www.suez.com/en/careers

Veolia
www.veolia.com/en/careers

FOR UPDATED AND ADDITIONAL RESOURCES, INCLUDING EXTRA CHAPTERS, GO TO WWW.ENV.CAREERS

References

All Party Parliamentary Sustainable Resource Group, UK (2012). 'Sustainable skills: the future of the waste management industry. https://www.policyconnect.org.uk/research/report-sustainable-skills-future-waste-management-sector (accessed 14 April 2021).

Allied Analytics LLP (2019). Waste management market by waste type and service: global opportunity analysis and industry forecast, 2018–2025.

CIWEM (2020). Contaminated land network. www.ciwem.org/networks/contaminated-land (accessed 18 September 2020).

Prospects (2020). Job profile: waste management officer. www.prospects.ac.uk/job-profiles/waste-management-officer (accessed 15 September 2020).

Pulp and Paper Technology (2021). Top 10 waste management companies in the world. www.pulpandpaper-technology.com/articles/top-10-waste-management-companies-in-the-world (accessed 14 April 2021).

UNEP (2015). Global waste management outlook. www.unenvironment.org/resources/report/global-waste-management-outlook (accessed 14 April 2021).

Waste Atlas Partnership (2014). Waste atlas the world's 50 biggest dumpsites. www.atlas.d--waste.com/Documents/Waste-Atlas-report-2014-webEdition.pdf (accessed 14 April 2021).

Waste Recruit Blog (2020). Innovations shaping waste and resource management. www.wasterecruit.com/news-blog/16/innovations-shaping-waste-and-resource-management (accessed 18 September 2020).

13

Renewables and Energy

13.1 Sector Outline

Globally, there has been significant growth in renewables and 'green' energy production in the past decade. However, this growth has been very patchy in terms of long-term government commitment to funding in some regions.

This is a very diverse sector with careers in many disciplines, from engineering roles to project management, design, environmental assessment and wider consultancy.

The International Energy Agency (IEA) noted, in its 2019 Global Energy Review:

> The year-on-year growth of renewables generation was 6.5%, faster than any other fuel including coal and natural gas. The share of renewables in global electricity supply reached 27% in 2019, the highest level ever recorded. Wind power, solar PV and hydropower together made up over 85% of renewables growth, complemented mainly by bioenergy.

Wind power increased by about 150 TWh year-on-year, the most of any renewable power generation technology, raising its share of electricity supply from 4.7% to 5.2% in 2019. The European Union, China and United States led wind output growth with a combination of offshore and onshore projects coming online and favourable weather conditions.

Solar PV electricity generation increased by about 130 TWh globally in 2019, second only to wind in absolute terms, reaching 2.7% of electricity supply. Solar PV's year-on-year growth of 22% far exceeded that of wind power, though this growth was significantly lower compared to 2018. Meanwhile, the European Union, India and the United States contributed similarly to the solar output increase. Solar PV accounts for almost 3% of global electricity mix.

Hydropower contributed over 100 TWh to the 2019 global increase in renewables generation, lifting its share of electricity supply to 16% and remaining the largest source of renewable electricity. In China, Brazil and India, hydropower increased more in 2019 in absolute terms than the ten-year average due to water availability and new hydropower projects. Hydropower will continue to play a key role in clean energy transitions by providing cost-effective low-carbon electricity and flexibility services that improve the reliability of power systems.

Global Environmental Careers: The Worldwide Green Jobs Resource, First Edition. Justin Taberham.
© 2022 John Wiley & Sons Ltd. Published 2022 by John Wiley & Sons Ltd.

Electricity generation from bioenergy rose 8%, maintaining its global share of electricity supply at about 2.5%. Growth was mainly driven by new projects in China, thanks to the country's policy target of 23 GW by 2020.

Source: IEA Global Energy Review 2019.

There are key issues and opportunities globally for renewable energy to exploit.

The World Bank (2017) 'State of Electricity Access Report 2017' outlined key global messages in access to electricity:

- Given current conditions, universal electricity access will not be met by 2030 unless urgent measures are taken. While nearly 1 billion people in Sub Saharan Africa alone may gain electricity access by 2040, due to population growth, an estimated 530 million people in the region will not have electricity access (IEA 2014).
- This energy shortfall must be rectified if the international community hopes to meet the 2030 Sustainable Development Goals, in light of the linkages between energy and other sustainable development challenges – notably, health, education, food security, gender equality, poverty reduction, and climate change.
- In many countries with low levels of electrification access, both grid and off-grid solutions are vital for achieving universal electricity access – but they must be supported by an enabling environment with the right policies, institutions, strategic planning, regulations, and incentives.
- Against a backdrop of climate change, plummeting costs for renewable energy technologies and adequate energy efficiency measures offer a tremendous opportunity for countries to be creative about electricity access expansion – with the emphasis on 'clean energy'.
- Emerging and innovative energy service delivery models offer unprecedented opportunities for private sector-driven off-grid electrification and accelerating universal electricity access – but only if countries can create the necessary environment for them to be replicated and scaled up.

The renewable energy sector has experienced significant global growth in recent decades. As noted by Deloitte in their 2018 Deloitte Insight 'Global Renewable Energy Trends' article, renewables are becoming a 'preferred' energy choice globally. Deloitte's report notes seven 'enabling trends' and 'demand trends' driving this process:

Enabling Trends

1) Renewables are reaching price and performance parity on the grid and at the socket. At one stage, renewables were seen as a 'niche' choice, but recently solar and wind can undercut the price of other energy sources.
2) Solar and wind can cost-effectively help balance the grid. The perception that integration of solar and wind into the grid has now been demonstrated as false. Storage technologies for energy are also helping the development of the solar and wind sector.
3) New technologies are honing the competitive edge of wind and solar. Rather than wait for innovation from other sectors, the renewables sector has forged ahead with its own developments and therefore is leading other sectors.

Demand Trends

4) Smart Renewable Cities are increasingly depending on solar and wind power to be a key energy source. Greenfield sites have no legacy hang-ups and generally want to demonstrate 'future proof' technologies.

5) Community energy on and off the grid. The growth in community energy delivery is aided by flexible systems, normally based on solar and wind energy. In less-developed countries, community energy delivery is cheaper than other options and more reliable.

6) Emerging markets as leading markets. In 2013, onshore wind growth in the less-developed world passed that of the developed world. This pattern is being replicated for other renewables. This shift in the global market has led to emerging markets being dominant.

7) The growing scope of corporate involvement. Corporates are able to purchase energy with contracts using Power Purchase Agreements (PPAs) and other processes. The procurement process for this often includes renewables targets in a process called 'additionality'. The standardisation for the procurement process and the compliance this ensures enables renewables to develop with an established market and surety of a binding agreement.

Renewable energy offers a very good opportunity for jobs to migrate across to the green sector from traditional heavy industry roles. The sector has roles in areas including manufacturing, site installation, servicing and engineering design as well as research, environmental management and impact assessment. This 'role transferability' makes it an attractive choice for governments globally as the transition process for jobs, skills and infrastructure is streamlined compared to implementing very 'different' new technologies.

PPAs are growing in popularity. As reported in The Guardian (2020):

> The world's biggest tech companies fuelled a record surge in the amount of renewable energy sold directly to global corporations last year, according to new figures.

> The amount of clean energy from renewable energy developers bought by companies has tripled in the past two years, driven by a growing corporate appetite for sustainable energy. The so-called power purchase agreements (PPAs) are likely to have cost between $20 billion and $30 billion (£15 billion to £23 billion), making up about 10% of the world's total renewable energy investments in 2019.

The market for PPAs has grown rapidly in North America, where most such purchases are made, but a growing number of deals were struck in Europe and Latin America.

13.2   Issues and Trends

Global Trends

UN Environment Programme (UNEP)'s 9th annual 'Global Trends in Renewable Energy Investment' report, prepared by Frankfurt School – UNEP Centre and Bloomberg New Energy Finance March 2015 noted that in 2014 global green energy investment had surged by 17% to $270 billion despite much lower oil prices. The leading renewable energies were solar and wind, accounting for 92% of overall investment. The amount of energy production added worldwide was approximately that of all US nuclear plants. The key investments in the period were solar expansion in China and Japan and offshore wind investment globally. There has been a corresponding redline in technology costs over time, but most markedly in solar and wind. Wind, solar, biomass and waste-to-power, geothermal, small hydro and marine power contributed to an estimated 9.1% of world electricity generation in 2014, up from 8.5% in 2013. The key investors in renewable energy in 2014 were:

China – $83.3 billion – up 39% from 2013
United States – $38.3 billion – up 7%
Japan – $35.7 billion – up 10%

Overall investment in solar increased by 25% to $149.6 billion and wind was up 11% to $99.5 billion globally. In solar, China and Asia dominated investment with around half the annual total. Offshore wind in Europe boomed. Geothermal investment increased 23% to $2.7 billion invested.

However, certain sectors declined in investment, including biofuels (down 8% to $5.1 billion), biomass and waste to energy (down 10% to $8.4 billion) and small hydro (down 17% to $4.5 billion).

The key feature noted in the UNEP report was rapid expansion of renewables in new markets in developing countries such as China, Brazil, India and South Africa. More than $1 billion was invested in Indonesia, Chile, Mexico, Kenya and Turkey.

There are challenges in the renewable energy sector in terms of policy uncertainty and structural issues in the electricity system in terms of handling changing energy mixes. Also, changing world oil prices have had an impact on decisions made globally in energy production options. Constantly changing government support schemes for renewables also can dent investor confidence.

In its report 'Renewable Energy and Jobs – Annual Review 2017', the International Renewable Energy Agency (IRENA) (2017) noted that more than 9.8 million people were employed globally in the renewable energy sector, up from just over 7 million in 2012. Jobs in the solar and wind sectors had doubled over the same period. There was an estimation that by 2030 there could be 24 million in the sector, more than offsetting potential losses in fossil fuel jobs. The leading countries for renewables were China, Brazil, United States, India, Japan and Germany. Solar photovoltaic (PV) was the largest employer in 2016 with leading countries China, United States and India. The largest renewable energy sector by installed capacity was large hydropower, and it also employed 1.5 million worldwide.

Asia was the key sector for renewables jobs, with 62% of the global total. In terms of manufacturing and installation, jobs have shifted to Asia, and Malaysia and Thailand have become global centres for solar PV fabrication.

China

An article by Patrick Caughill in Futurism magazine in 2018 highlighted China's expansion to become a dominant world force in renewable energy. The Institute for Energy Economics and Financial Analysis (IEEFA) in 2018 their report 'China 2017 Review: World's Second-Biggest Economy Continues to Drive Global Trends in Energy Investment' stated that in 2017 China's total investment in clean-energy projects represented more than $44 billion in investment – a significant growth from 2016's $32 billion. The US decision to withdraw from the Paris Climate Agreement was a catalyst for China to accelerate its growth in renewable energy. Tim Buckley, the IEEFA's Director of Energy Finance Studies noted:

Although China isn't necessarily intending to fill the climate leadership void left by the U.S. withdrawal from Paris, it will certainly be very comfortable providing technology leadership and financial capacity so as to dominate fast-growing sectors such as solar energy, electric vehicles, and batteries.

In 2018, Helen Davidson in The Guardian (United Kingdom) reported that China is moving towards becoming a global leader in renewable technology and that their trajectory stands in contrast to the United States. China's Belt and Road agenda was also a stimulus for China's growth in the sector. So far, China has exported $8 billion of solar equipment and is now the number one exporter of environmental goods and services.

The IEEFA report noted:

> China's presence in wind power globally is also on the rise, led by international activities of companies such as Goldwind and by China Three Gorges' diversification away from hydroelectricity. . .That allows it to further project itself globally as a responsible major power while addressing its domestic air pollution concerns and building world-leading capacity in new energy markets.

In August 2017, China's top coal mining company, Shenhua Group Corp., merged with one of the 'big five' power utilities, China Guodian Corp. The newly named China Energy Investment Corp. created the world's largest power generator, and with the clean-energy assets of Guodian, Shenhua's growth was no longer dependent on the pursuit of coal.

The IEEFA also reported that China was 'outmaneuvering other economies' in securing energy commodity supplies such as lithium, nickel and cobalt, allowing them to dominate manufacturing of batteries and electric vehicles.

A McKinsey Global Institute report from 2017 'Beyond the Supercycle: How Technology Is Reshaping Resources' noted that China announced that it would invest $360 billion in renewable energy by 2020 and scrap plans to build 85 coal-fired power plants. Chinese authorities also reported that the country was already exceeding official targets for energy efficiency, carbon intensity and the share of clean energy sources.

The indication is that China aims to phase out coal and transition to renewables and other energy sources. Also, as China develops into a service and consumption economy, it can influence the global resource sector. China is investing £32 billion in renewables overseas and also is developing its manufacturing in wind and solar to enable it to internally source its own needs and sell to the global markets.

Africa

An article by Kurt Davis Jr. in the media outlet Africa.com noted in 2014:

> An energy deficit has effectively stunted Africa's development, with an estimated 70 percent of people in sub-Saharan Africans without reliable access to electricity. . . Fossil fuel-based power generation is the most expensive form of energy globally, yet it is the largest source of electricity generation in Africa. This is the least of concern for a continent simply trying to boost its total power capacity from the current 147 GW, according to the African Development Bank, which is equivalent to the total capacity installed in Belgium, and what China installs every one to two years.

The IEA noted in 2014 that sub-Saharan Africa will require more than $30 billion in investment to achieve universal electricity by 2030. Rural sub-Saharan Africa will require the vast amount of the funds, with more than 85% of those living in rural areas lacking access to reliable electricity.

There are considerable challenges but also opportunities for renewable energy in Africa. Key challenges are a lack of investment, both rural and urban, and an over-focus on basic electrical supply needs from whatever source is available. Opportunities are that concerns over global emissions provide an investment opportunity in African green power generation.

Hydropower provides a great opportunity across the entire sub-Saharan African region, an example being The Grand Renaissance Dam in Ethiopia which is expected to deliver up to 6,000 MW to the country, with neighbouring Djibouti and Somalia inquiring about the potential of importing such energy. The dam started filling in July 2020, and it is also expected to create up to 12,000 jobs, as well as providing services in initiation investments in the likes of agriculture and fishing.

Sub-Saharan African wind production is booming, with East Africa seeing a major bump in wind energy generation. Wind energy commitments in Kenya skyrocketed from zero in 2011 to $1.1 billion in 2012, underscored by the Lake Turkana Wind power project which will provide 300 MW to the Kenya electrical grid.

Geothermal energy continues to emerge as a potential hidden gem in the sub-Saharan African electricity grid. Recent projects in Ethiopia, Rwanda, Ghana and Nigeria speak to the potential and rising interest as geothermal opportunities are also related to the emerging gas and oil discoveries. A $4 billion geothermal plant project in Ethiopia noted in the Industry Week in 2013 could set the standard for projects in the region.

Coal and gas power plants will continue to grab headlines in Africa. New gas discoveries in Tanzania and Mozambique as well as oil booms in Angola and Ghana should not overshadow Africa's enthusiastic efforts to develop renewable energy.

The Planet Earth Institute, an international NGO and charity working for the scientific independence of Africa, in 2017 reported:

> Sub-Saharan Africa has access to a wealth of primary renewable energy supplies, with enough geothermal, hydro, wind and solar resources to provide terawatts of power. The continent has the potential to source an additional 10 terawatts of solar energy, 1,300 gigawatts of wind power, and 1 gigawatts of geothermal potential... By successfully taking advantage of its abundant clean power resources, Africa has the potential to propel itself towards a renewable energy revolution. Through this, the continent could see large economic growth, the creation of a thriving job market and, importantly, improved quality of life for millions of people.

North America

Deloitte in its 2020 Renewable Energy Industry Outlook: A Midyear Update highlighted:

> For the first time ever, in April 2019, renewable energy outpaced coal by providing 23 percent of US power generation, compared to coal's 20 percent share. In the first half of 2019, wind and solar together accounted for approximately 50 percent of total US renewable electricity generation, displacing hydroelectric power's

dominance. . .Currently, in many areas of the United States, the levelized cost of energy for onshore wind is less than for utility solar photovoltaic (PV). However, costs for solar have been declining faster than wind recently. . .solar could become increasingly cost-competitive with wind. As a result, in 2020 and beyond, some wind-only customers will likely diversify and build a mixed portfolio of wind, solar, and storage to fulfill commitments.

The year ahead promises further growth in the renewable energy sector. This will likely come against a backdrop of increased innovation and collaboration among multiple stakeholders. Renewables are likely to continue moving into the driver's seat in electricity markets as utilities and regulators prefer them to replace retiring capacity and customers increasingly choose them to save costs and address climate change concerns. Growth in the US offshore wind sector will likely bring multiple opportunities for industry players as states vie for manufacturing and port infrastructure projects. Grid resiliency will also likely be a growing driver for distributed renewable deployment as utilities and their customers increasingly consider renewable microgrids combined with storage solutions. However, trade and tariff policy uncertainty will likely keep the industry on the lookout for risk mitigation tactics. But companies that are ready to innovate, collaborate, and seize new opportunities will likely thrive in a new phase of renewable growth.

Europe

According to Eurostat (2020):

> Becoming the world's first climate-neutral continent by 2050 is the objective behind the European Green Deal, the most ambitious package of measures that should enable European citizens and businesses to benefit from sustainable green transition.

The growth in electricity generated from renewable energy sources during the period 2008 to 2018 largely reflects an expansion in three renewable energy sources across the EU, principally wind power, but also solar power and solid biofuels (including renewable wastes). In 2018 wind power is the single largest source for renewable electricity generation in the EU. Indeed, the amount of electricity generated from hydro was relatively similar to the level recorded a decade earlier. By contrast, the quantity of electricity generated in the EU from solar and from wind turbines was 15.5 times and 2.9 times as high in 2018 as it had been in 2008. The growth in electricity from solar power has been dramatic, rising from just 7.4 TWh in 2008 to 115.0 TWh in 2018.

13.3 Key Organisations and Employers

The IRENA (2018) reported that globally there were 10.3 million jobs in renewables, 43% of which were in China, with most employment focused in a small number of countries including China, Brazil, United States, India, Germany and Japan.

Thomson Reuters (2017), in its report '2017 Top 100 Global Energy Leaders', highlighted the following companies as leading in the renewables sector:

Canadian Solar Inc. – Canada
CropEnergies – Germany
First Solar – United States
GCL-Poly Energy Holdings Ltd. – Hong Kong
Global Pvq SE i l – Germany
Green Plains Inc. – United States
Guodian Technology & Environment Group Corp. Ltd. – China
Hanergy Thin Film Power Group Ltd. – Hong Kong
Inox Wind – India
Jiangsu Akcome Science & Technology Co., Ltd. – China
Motech Industries Inc. – Taiwan
Pacific Ethanol – United States
Renewable Energy Group Inc. – United States
Risen Energy Co., Ltd. – China
Shanghai Aerospace Automobile Electromechanical Co., Ltd. – China
Siemens Gamesa Renewable Energy – Spain
SolarWorld Industries – Germany
SunEdison – United States
Sungrow – China
SunPower Corp. – United States
Suzlon Energy Ltd. – India
TPI Composites – United States
VERBIO Vereinigte BioEnergie – Germany
Vestas – Denmark
Xiangtan Electric Manufacturing Co., Ltd. – China

Manufacturers account for a high proportion of jobs, with global manufacturing very focused in a few countries. In sectors such as large hydropower, operation and maintenance work has the greatest employment.

Wind

According to NS Energy (2019):

> The world's overall capacity of installed wind power reached 600 gigawatts (GW) in 2018, courtesy of the booming production rates of some of its top wind turbine manufacturers. According to GlobalData, Danish firm Vestas re-established its place at the top of the list, followed by China's Goldwind, Siemens Gamesa, GE Renewable

Energy and Envision Energy. . . The top five turbine manufacturers accounted for nearly 62% of the total installed capacity in 2018, up almost 5% on the previous year

The comprehensive report 'Careers in Wind Energy' by the US Bureau of Labor Statistics in 2010 outlines career opportunities in the United States across the different elements of the wind power sector. It notes:

> Wind-generating capacity in the United States grew 39 percent per year from 2004 to 2009 and is expected to grow more rapidly as demand for renewable energy increases. . .it will provide many opportunities for workers in search of new careers. These careers extend beyond the wind farm: it also takes the efforts of workers in factories and offices to build and operate a turbine. . .an estimated 85,000 Americans are currently employed in the wind power industry and related fields. . .wind power is only a tiny segment of the national energy market. In 2009, wind energy made up 1.8 percent of U.S. power generation, an increase from 1.3 percent in 2008. However, wind power accounts for about 50 percent of renewable energy.

In 2019, according to the American Wind Energy Association (2021), wind power generated 7.2% of the country's electricity which is a significant expansion.

Solar

The market research company Technavio in 2019 listed their top solar companies in the world as:

JinkoSolar – China
Canadian Solar – Canada
Trina Solar – China
SunPower Corp. – United States
Hanwha Q Cells – South Korea
JA Solar – China
LONGi Solar – China
Risen Energy – China
GCL-SI – Hong Kong
Talesun – China

They noted 'Solar photovoltaic power (PV) is one of the fastest growing segments of the global renewable power generation in 2017, as deployment boomed in China and prices continued to fall.'

Hydroelectric

NS Energy noted in 2019:

> 'At 1,295 GW, hydropower accounted for approximately 55% of the global installed renewable energy capacity in 2018. China, Canada, Brazil, Russia, and France host some of the world's biggest hydropower producers.'

Based on installed capacities, the leading companies were:

China Yangtze Power Co. – China

Centrais Eletricas Brasileiras (Eletrobras) – Brazil

Hydro-Quebec – Canada

RusHydro – Russia

Electricite de France (EDF) – France

Geothermal

PR Newswire in 2019 highlighted the leading companies in Geothermal as:

Turboden S.p.A. – Italy

Green Energy Geothermal – Iceland (acquired by the Canadian company Energy
Co-Invest Global Corp. (ECC Global) in 2018)

Berkshire Hathaway Inc. – USA

Terra Gen – USA

Reykjavik Geothermal – Iceland

Alterra Power Corporation – Canada

TAS Energy – USA

Atlas Copco Group – Sweden

Exergy – Italy (acquired by China's Nanjing TICA Thermal Technology Co. Ltd in 2019)

KenGen – Kenya

Halliburton – USA

Enel Green Power – Italy

ElectraTherm – USA (acquired by BITZER in 2016)

Calpine Corporation – USA

Fuji Electric Co. Ltd. – Japan

Toshiba Corporation – Japan

Mitsubishi Heavy Industries – Japan

General Electric – USA

Ansaldo Energia – Italy

Ormat Technologies – USA

13.4 Careers in the Sector

Across the renewables and energy sector, different parts of the 'process' offer different careers:

- Research and design
- Regulation and legislation
- Manufacturing

- Project planning and management
- Operation and maintenance

Roles within the process include:

- Engineers and technicians work across the sector in all phases – they often have transferred skills from other energy roles
- Servicing and maintenance technicians during operations
- Research and testing specialists in areas including materials, design, civil, electrical and mechanical engineering, hydraulics, aeronautics, flow specialists and atmospheric scientists
- Project managers and planners
- Community liaison specialists
- Policy and legal advisors
- Manufacturers, quality controllers and technicians in what is known as the 'Craft' sector

13.5 Job Titles in the Sector

There are many roles within the renewables and energy sector. The US Bureau of Labor Statistics (2010) separates these into Manufacturing, Project Development and Maintenance and Operation.

Manufacturing

Mechanical engineer, controls specialist, design engineer, electrical and instrument engineer, Heating, Ventilation, and Air Conditioning (HVAC) designer, load analyst, plant engineer, systems engineer, technical analyst

Project Development

Building services engineer, electrical engineer, civil engineer, commissioning manager, project manager, energy manager, equipment engineer, facilities coordinator, hydrogeologists, installation manager, project coordinator, purchasing specialist, sustainability engineer

Operation and Maintenance

Energy auditor, environmental manager, carbon consultant, energy manager, health and safety manager, maintenance manager, technician

13.6 Educational Requirements

Within this diversifying and rapidly developing sector there are differing requirements for roles:

- There are many 'Craft' roles which require relevant skills and experience.
- Engineers generally have at least a bachelor's degree in an engineering field.
- Industry certifications may be expected for some roles.
- A significant number of jobs seek higher-level degrees as well as experience.
- Research roles will often require PhD level qualifications.

- Experience in the sector applied for is often deemed essential.
- On the job training is often given before assignment to a site and during employment.
- Internships and relevant work experience are valuable.

13.7   Personal Attributes and Skill Sets

Helpful skills and attributes may include:

- Teamwork
- Communications skills
- Technical report writing
- High level IT skills, often in Geographic Information System (GIS), modeling and data analysis
- Project and budget management
- Knowledge of industry trends, legislation, work practices and regulations
- The ability to get technical points across to different audiences

Training

European Energy Centre
www.euenergycentre.org/training
www.euenergycentre.org/training/usa-market-trends-and-finance-course

Renewableenergyjobs.com 'Complete Guide to Renewable Energy Training and Education'
www.renewableenergyjobs.com/content/complete-guide-to-renewable-energy-training-
    and-education

Narec Distributed Energy
www.narecde.co.uk/funded-renewable-energy-training-3

powerEDGE Asia
www.poweredgeasia.com

Energy Institute
www.energyinst.org/whats-on/training

Globally delivered and online training
Green Power Academy
www.greenpoweracademy.com

Kellie Flanagan, Renew magazine (2018) 'Renewable energy courses guide'
www.renew.org.au/renew-magazine/buyers-guides/renewable-energy-courses-guide

Australian Government 'Energy efficiency training'
www.energy.gov.au/business/energy-efficiency-skills-and-training/
    energy-efficiency-training

IRENA 'Regional Model Analysis and Planning Programme'
www.irena.org/energytransition/Energy-Planning-Support/Regional-training-workshops

United Nations Industrial Development Organization (UNIDO) (2016) '150,000 students
    in Latin America and the Caribbean take online courses in renewable energy'
www.unido.org/news/150000-students-latin-america-and-caribbean-take-online-courses-
    renewable-energy

13.8   Career Paths and Case Studies

◀ **Top Tips**

David Brown, Head of Energy and Sustainability, ISS, United Kingdom

1) Most recruiters want experience – get relevant work experience anyway you can, e.g. study related, internship and volunteering.
2) Network, network – LinkedIn allows you to find companies and people in your industry effortlessly.
3) Create a work portfolio.
4) Really understand what a recruiter wants.
5) Find a mentor in the industry.

Source: David Brown, Head of Energy and Sustainability, ISS, UK. © John Wiley & Sons.

◀ **Top Tips**

Sydney Wang – Founder of Green Career Path, Research Analyst at a US Energy Research Institute, United States

1) Be a life-long learner. The environmental industry, like any other industry in the 21st century, has been in need of talented people with an interdisciplinary background and a diverse skillset. What we learned in the environmental or engineering schools may not be enough for us to solve real-world problems. No matter if you are a recent graduate looking forward to jumpstarting your career, or an industry expert with 30 years of experience, keeping an open mind and always being ready to learn new things are essential to your career development in the environmental industry. From my experience as an ex-consultant, an energy researcher and a social entrepreneur in both the United States and China, I have already seen great opportunities in the intersection of environment, data science, urban planning, international development, social justice and many more.
2) Build your own professional network now. People have always been talking about the importance of networking for getting a job, but it is too late if you start networking right before you need a new job. Networking actually has never been easier when all the conferences and events are virtual due to the global pandemic now. You don't even need to grab a coffee to talk to someone. Also, networking to me is more about connecting to people of similar interests and staying informed in this ever-changing world, while my next dream job is probably a bonus.

Personal Profile

After studying in both the United States and China in the fields of environmental science and policy, Sydney worked in a global-leading consulting firm and provided mining, aircraft, pharmaceutical and manufacturing companies with enterprise-level Environment, Health and Safety and Sustainability (EHS&S) software and data management solutions that are tailored to their needs. Then she joined an energy research institute to conduct research on energy consumption and Greenhouse Gas (GHG) emissions modelling.

She also started Green Career Path, an initiative that helps students and young professionals explore the environment, energy and sustainability fields and develop their career by delivering a series of small-group seminars on new technologies, connecting people, sharing daily new job posts and providing mentorship.

www.linkedin.com/company/sustainability-career-path/about

Source: Sydney Wang - Founder of Green Career Path, Research Analyst at a US Energy Research Institute, USA. © John Wiley & Sons.

◀ **Top Tips**

James Beal, Specialist, Department for International Trade, United Kingdom

I think the top tips for the sector have fundamentally changed over the years. In the start it was knowledge and passion, probably verging in belief that you can be part of the change.

Basically, you needed something extra to overcome the forces of climate change deniers, the people fearing change and those set on continuing as we always have. Now that battle is largely solved, and the low-carbon agenda is mainstream – and in many ways not really a sector in itself – it's business sense.

However, for people my age the challenge within the sector is that so many 'passionate' people remain in the sector that pragmatism can't be relied on to resolve differences. Which often stifles success.

Which brings me back to the start – that what was needed isn't now – it's big business and those able to manage and believe within big business will succeed. Work in a team. Priority management. Good communicator. Deliver tasks.

Source: James Beal, Specialist, Department for International Trade, UK. © John Wiley & Sons.

13.9 External Resources

United Kingdom

Energy UK
www.energy-uk.org.uk

Jillian Ambrose, The Guardian (2020) 'Renewable energy breaks UK record in first quarter of 2020'
www.theguardian.com/business/2020/jun/25/renewable-energy-breaks-uk-record-in-first-quarter-of-2020

Renewable Energy Jobs
www.renewableenergyjobs.com

Green Jobs
www.greenjobs.co.uk

Prospects (2020), Renewable Energy Careers
www.prospects.ac.uk/jobs-and-work-experience/job-sectors/energy-and-utilities/renewable-energy-careers

REA
www.r-e-a.net

Energy Saving Trust
www.energysavingtrust.org.uk/renewable-energy

Europe

European Commission 'Renewable energy'
www.ec.europa.eu/energy/topics/renewable-energy_en

BUILD UP (2020) 'In focus: Renewable energy in Europe'
www.buildup.eu/en/news/focus-renewable-energy-europe

RES LEGAL Europe (2019) 'Legal sources on renewable energy'
www.res-legal.eu

Eurostat (2020) 'Renewable energy statistics'
www.ec.europa.eu/eurostat/statistics-explained/index.php/Renewable_energy_statistics

Asia

IRENA Asia and Pacific
www.irena.org/asiapacific

Asian Development Bank (2020) '12 things to know: The rise of renewable energy in Asia
and the Pacific'
www.adb.org/news/features/12-things-know-rise-renewable-energy-asia-and-pacific

PwC Singapore (2018) 'The next frontier of infrastructure investments – Renewable energy
in Asia-Pacific'
www.pwc.com/sg/en/publications/renewable-energy-in-asia-pacific.html

The ASEAN Post (2020) 'Renewable energy challenges in Southeast Asia'
www.theaseanpost.com/article/renewable-energy-challenges-southeast-asia

Africa

IRENA – Africa
www.irena.org/africa

Africa 2030: Roadmap for a renewable energy future
www.irena.org/publications/2015/Oct/Africa-2030-Roadmap-for-a-Renewable-Energy-
Future

IEA (2019) 'Africa energy outlook 2019'
www.iea.org/reports/africa-energy-outlook-2019

Renewable Energy Africa
www.renewable-energy-africa.com

North America

C2ES – the Center for Climate and Energy Solutions
www.c2es.org/content/renewable-energy

U.S. Energy Information Administration (EIA) 'Renewable energy explained'
www.eia.gov/energyexplained/renewable-sources

Government of Canada 'About renewable energy'
www.nrcan.gc.ca/energy/energy-sources-distribution/renewables/about-renewable-
energy/7295

The Women's Council on Energy and the Environment (WCEE) focuses on women, energy
and the environment
www.wcee.org

South America

NS Energy (2020) 'Top five renewable energy producers of South America'
www.nsenergybusiness.com/features/renewable-energy-producers-south-america

IRENA (2016) 'Renewable energy market analysis: Latin America'
www.irena.org/publications/2016/Nov/Renewable-Energy-Market-Analysis-Latin-
America

Energy Voice (2020) 'Latin America's renewable energy capacity set to skyrocket to 123 GW
by 2025'
www.energyvoice.com/oilandgas/americas/264127/latin-america-renewables-capacity-
rystad

Oceania

Australian Renewable Energy Agency (ARENA)
www.arena.gov.au

Adam Morton, The Guardian (2020) 'With record new solar and wind installed, Australia's
clean energy is booming – for now'
www.theguardian.com/australia-news/2020/sep/06/with-record-new-solar-and-wind-installed-
australias-clean-energy-is-booming-for-now

Energy Australia 'Supporting renewable energy'
www.energyaustralia.com.au/about-us/sustainability/clean-energy-transition/supporting-
renewable-energy

Clean Energy Council (2020) 'Clean energy Australia report'
www.cleanenergycouncil.org.au/resources/resources-hub/clean-energy-australia-report

Ministry of Business, Innovation & Employment (2020) 'Energy in New Zealand 2020'
www.mbie.govt.nz/dmsdocument/11679-energy-in-new-zealand-2020

Global

International Renewable Energy Agency (IRENA)
www.irena.org/events

The Energy Institute
www.energyinst.org/about

Matchtech
www.matchtech.com/renewable-energy-jobs

Energy Jobline
www.energyjobline.com/renewables

Shaw Energy
www.shawenergyltd.com

Renewable Energy Jobs (2009–2021)
www.renewableenergyjobs.com/content/a-z-of-renewable-energy-jobs-green-collar-
  jobs-titles

Green Recruitment Company
www.greenrecruitmentcompany.com

**FOR UPDATED AND ADDITIONAL RESOURCES, INCLUDING EXTRA CHAPTERS, GO TO
WWW.ENV.CAREERS**

References

Ambrose, J., The Guardian UK (2020) 'Tech giants power record surge in renewable energy
  sales' www.theguardian.com/environment/2020/jan/28/google-tech-giants-spark-record-
  rise-in-sales-of-renewable-energy (accessed 18 September 2020).
American Wind Energy Association (2021)'Wind facts at a glance' www.awea.org/wind-101/
  basics-of-wind-energy/wind-facts-at-a-glance (accessed 14 April 2021).
Caughill, P. (2018) Futurism magazine 'China Is the New World Leader in Renewable Energy'
  www.futurism.com/china-new-world-leader-renewable-energy (accessed 14 April 2021).
Davidson, H. (2018) The Guardian (UK) 'China on track to lead in renewables as US retreats,
  report says' www.theguardian.com/environment/2018/jan/10/china-on-track-to-lead-in-
  renewables-as-us-retreats-report-says (accessed 14 April 2021).
Davis Jr, K.,(2014) Africa.com 'Africa's Renewable Energy Potential' www.africa.com/africas-
  renewable-energy-potential/ (accessed 14 April 2021).
Deloitte (2018) Deloitte Insight 'Global Renewable Energy Trends' www2.deloitte.com/global/
  en/pages/energy-and-resources/articles/global-renewable-energy-trends.html (accessed
  19 September 2020).
Eurostat (2020) Renewable Energy Statistics www.ec.europa.eu/eurostat/statistics-explained/
  index.php/Renewable_energy_statistics#Share_of_renewable_energy_almost_doubled_
  between_2004_and_2018 (accessed 18 September 2020).
Hamilton, J. and Liming, D. (2010) Bureau of Labor Statistics, USA 'Careers in Wind Energy'
  www.bls.gov/green/wind_energy/ (accessed 14 April 2021).

IEA Global Energy Review 2019 www.iea.org/reports/global-energy-review-2019/renewables# abstract (accessed 18 September 2020).

Industry Week (2013) 'Ethiopia Signs $4 Billion Geothermal Deal' www.industryweek.com/ operations/energy-management/article/21961475/ethiopia-signs-4-billion-geothermal-deal (accessed 18 September 2020).

Institute for Energy Economics and Financial Analysis (IEEFA) (2018) 'China 2017 Review: World's Second-Biggest Economy Continues to Drive Global Trends in Energy Investment' www.ieefa.org/ieefa-report-china-continues-position-global-clean-energy-dominance-2017/ (accessed 14 April 2021).

International Renewable Energy Agency (2018) (IRENA) Annual Review 2018 'Renewable Energy and Jobs' www.irena.org/-/media/Files/IRENA/Agency/Publication/2018/May/ IRENA_RE_Jobs_Annual_Review_2018.pdf (accessed 18 September 2020).

IRENA (2017) 'Renewable Energy and Jobs – Annual Review 2017' www.irena.org/ publications/2017/May/Renewable-Energy-and-Jobs--Annual-Review-2017 (accessed 14 April 2021).

McKinsey Global Institute (2017). Beyond the Supercycle: How Technology is Reshaping Resources.

NS Energy (2019) 'What are the largest hydropower companies in the world?' www. nsenergybusiness.com/features/largest-hydropower-companies/ (accessed 14 April 2021).

PR Newswire (2019) Top 20 Geothermal Power Companies 2019 www.prnewswire.com/ news-releases/top-20-geothermal-power-companies-2019-300860920.html (accessed 14 April 2021).

Planet Earth Institute (2017) 'The Future of Renewable Energy in Africa: Challenges and Opportunities' www.planetearthinstitute.org.uk/wp-content/uploads/2017/10/Spotlight-Seminar-Renewable-Energy.pdf (accessed 14 April 2021).

Prospects (2020) Renewable Energy Careers www.prospects.ac.uk/jobs-and-work-experience/ job-sectors/energy-and-utilities/careers-in-the-renewable-energy-industry (accessed 14 April 2021).

Renewable Energy Jobs website(2009–2021) www.renewableenergyjobs.com/ content/a-z-of-renewable-energy-jobs-green-collar-jobs-titles

Technavio (2019) 'Top 10 Solar Energy Companies in the World 2019' https://blog.technavio. com/blog/top-10-solar-energy-companies (accessed 14 April 2021).

Thomson Reuters '2017 Top 100 Global Energy Leaders' www.thomsonreuters.com/content/ dam/ewp-m/documents/thomsonreuters/en/pdf/reports/thomson-reuters-top-100-global-energy-leaders-report.pdf (accessed 18 September 2020).

Todd, Felix in NS Energy (2019) 'Who were the world's top five wind turbine manufacturers in 2018?' www.nsenergybusiness.com/features/top-wind-turbine-manufacturers-2018/ (accessed 14 April 2021).

UNEP Centre and Bloomberg New Energy Finance (2015) UNEP's 9th annual 'Global Trends in Renewable Energy Investment' report, prepared by Frankfurt School www. unenvironment.org/resources/report/global-trends-renewable-energy-investment-2015 (accessed 18 September 2020).

World Bank (2017) 'State of Electricity Access Report 2017' www.worldbank.org/en/topic/ energy/publication/sear (accessed 14 April 2021).

14

Journalism and Writing/Blogging

14.1 Sector Outline

Within the sector there are a number of areas of work:

- Journalists – print and online
- Bloggers and vloggers
- Non-fiction Writers – in varied roles from book authors to corporate material writers – technical writers or communications/media materials
- Fiction writers
- Broadcast media employees

Environmentalscience.org (2020) notes:

> Environmental writers write about environmental topics for a variety of outlets. For example, they may cover energy, environmental policy, water issues, climate change, environmental justice, or new technologies and industry news. They may write in one style for one publication or employer, or many styles for different markets.

Many people in the sector have 'portfolio careers' where they work in a number of different areas for different clients such as a writer who could work as a magazine editor, green article writer, green blogger and corporate sustainability report writer within their portfolio.

In recent years, blogging and vlogging has become a distinct job option.

Full time roles for news and other organisations can be in roles such as writer, editor and journalist.

14.2 Issues and Trends

There has been a fundamental shift away from the print media towards broadcast media and new media. Newer platforms such as YouTube and the growth of the green media and online news services have enabled green writers to reach bigger audiences more swiftly.

Global Environmental Careers: The Worldwide Green Jobs Resource, First Edition. Justin Taberham.
© 2022 John Wiley & Sons Ltd. Published 2022 by John Wiley & Sons Ltd.

The shift to online media has largely been due to advertising and news viewing moving online. The Columbia Journalism Review (2016) in the USA noted:

> In 2005, for every one digital-only journalist, there were 20 newspaper journalists. In 2015, for every one digital-only journalist, there were four newspaper journalists.

The US Bureau of Labor Statistics (2016), in its report 'Employment trends in newspaper publishing and other media, 1990–2016' highlighted this change:

> Few industries have been affected by the digital or information age as much as newspapers and other traditional publishing industries (books, magazines, etc.). In June 1990, there were nearly 458,000 people employed in the newspaper publishing industry; by March 2016, that figure had fallen to about 183,000, a decline of almost 60 percent. Over the same period, employment in Internet publishing and broadcasting rose from about 30,000 to nearly 198,000.

Roy Greenslade of the Guardian in 2016 reported:

> Cutbacks have been the reality of the news trade in Britain for the best part of 20 years at national, regional and local level. And the same story has been unfolding in the USA, Canada, Australia and most European countries. The cause, as we all know, is the onward march of the digital revolution. We who praise its advance cannot also help but lament its disruption. . .no publisher, despite differing motivations, can escape the commercial effects of a technological revolution that is in the process of destroying the funding mechanism that has underpinned newspaper companies for more than 150 years.

Regional newspapers in the United Kingdom have shown significant declines, mainly fuelled by a growth in fully searchable advertisements by specialised online sellers. The Press Gazette (2017) noted that the rise of online advertising had led to a 70% cut in journalist numbers at larger UK regional newspapers.

The Reuters Institute for the Study of Journalism (2016) noted:

> Journalism plays a pivotal role in keeping us informed and critically aware. But in a period when digital communications technologies are violently disrupting news industry business models there is confusion and debate as to whether the result is less journalism, worse journalism or more and better journalism delivered through a more diverse array of media, including social media. . .
>
> The best paid jobs are still in television, where disruptive forces bearing on news are weaker. The proportion of journalists working in newspapers has fallen sharply, but disagreement about definitions makes it unsettled whether overall in the digital age we have more or less journalism and more or fewer journalists.
>
> The authors estimate that there are now 30,000 journalists working wholly or partly online, but many bloggers are excluded from this count, along with others whose journalistic identity is complex.

Digital influences also mean that journalists have more data about audience responses to their work; it remains unclear to what extent they feel bullied by this into the clickbait game, rather than feeling that they can use the data to make better, independent decisions about how to provide a service the audience values.

The report also stated that 2% of the reporters surveyed said they worked in the 'Environment' area.

It has to be noted that in many regions, environmental reporting is a dangerous career to develop. Every year, large numbers or journalists are injured and killed for their reporting.

The International Federation of Journalists (IFJ) (2018) reported that 81 journalists were killed in 2017 and that:

> unprecedented numbers of journalists were jailed, forced to flee, that self-censorship was widespread and that impunity for the killings, harassment, attacks and threats against independent journalism was running at epidemic levels.
>
> According to IFJ records, the Asia Pacific has the highest killing tally, followed by the Arab World and Middle East, The Americas, Africa then Europe.

14.3   Key Organisations and Employers

There are a number of key organisations in the sector. These include:

Society for Environmental Journalists
The Society of Environmental Journalists is the only North American membership association of professional journalists dedicated to more and better coverage of environment-related issues
www.sej.org

Association for the Study of Literature and Environment (ASLE)
ASLE seeks to inspire and promote intellectual work in the environmental humanities and arts
www.asle.org

The National Association of Science Writers
NASW is a community of journalists, authors, editors, producers, public information officers, students and people who write and produce material intended to inform the public about science, health, engineering, and technology
www.nasw.org

14.4   Careers in the Sector

Employers in the sector include:

- Newspapers and magazines
- Broadcast media

- Online news channels and websites
- Campaigning/NGO organisations
- Many people in the sector are self-employed and working with different clients
- An increasing number of bloggers, often on campaigning issues, build successful businesses from blogging and vlogging

14.5 Job Titles in the Sector

Editor, journalist, writer, staff writer, content developer.

14.6 Educational Requirements

- Many writers have a background in journalism and most countries have a structure for the journalism sector
- There are many journalism and English degree programmes
- At Masters level there are relevant degrees and there are short courses in the sector
- Bloggers and vloggers can have any number of backgrounds

14.7 Personal Attributes and Skill Sets

Skills and attributes which are helpful include:

- Passion for writing and/or broadcasting
- Well-developed researching and writing skills
- Writing to appeal to different audiences
- Ability to meet deadlines
- Discipline in writing
- Accuracy and ability to analyse then summarise information
- Technical writing and report writing
- Communication skills
- Project management

Training

Society of Environmental Journalists
www.sej.org/library/teaching-tools/overview

WFSJ and SciDev.Net 'Online Course in Science Journalism'
www.wfsj.org/course/en/index.html

Population Reference Bureau (2007) 'Reporting on Population, Health, and the Environment:
   A Guide for Central American Journalists'
https://assets.prb.org/pdf07/PHEMediaGuideCentralAmerica.pdf

Forum for African Investigative Reporters (FAIR) (2012) 'Investigating Environmental Issues for African Media Workers'

MediaClimate 'Education/Training'
www.mediaclimate.net/education-training

Society of Professional Journalists (2020) 'Journalists Toolbox – Environment Resources'
www.journaliststoolbox.org/2020/10/13/miscellaneous_environment_sites

14.8   Career Paths and Case Studies

Personal Profile

Erika Yarrow, Journalist, United Kingdom

Environmental journalism – What does it offer?
For inquiring individuals, with well-honed communication skills, journalism offers a diverse and rewarding career. To be in a position to source information from experts in their field; hold politicians to account; and challenge assumptions, is a great privilege.

A specialism enables a journalist to set themselves apart from the crowd and a specialism in the environment takes the privileged role of the journalist to a critically important place. The environment is not just an area where growth that is expected to increase in the coming decades as climate change continues to bite and increases in population and consumption put strains on resources and so business profits. More significantly, the environment is a 'sector' that touches all others. In fact, it touches everything. Be it business, politics, humanitarian relief, education or faith, there is a vital connection to the environment, because protection of the environment is not limited to the stereotypical, altruistic goal of tree hugging activists; it is critical to civilisation and necessary for communities to survive and thrive.

Hence, for a journalist, the environment offers great opportunities for exploration. Your research can take you anywhere in the world, enable you to learn from experts in an array of fields and provides the opportunity to challenge the status quo, shining light on injustice and hypocrisy.

As an environmental journalist you can help novel technologies gain recognition; hold stakeholders to account; and influence government and regulatory policy. If you are passionate about the environment and want to play a part in creating a future that nurtures the environment and those that depend upon it, a career in environmental journalism offers great opportunities.

The environment is a subject that will suck you in and hold you, touch every aspect of your life, and challenge you to consider new ways of living and thinking.

Source: Erika Yarrow, Journalist, UK. © John Wiley & Sons.

Personal Profile

Dr Mark Everard, Author, Broadcaster and Associate Professor, University of the West of England, United Kingdom

Journalism/writing/blogging as a green career

As a frequent broadcaster and writer – of books, magazine contributions, blogs and other media – I would firstly caution most newcomers about looking on these activities as a traditional career. Whilst mainstream journalism most definitely can be a career, much of the 'green media' depends on contributions (text and visuals) that are voluntary or paid largely in a token way. As an example, I have to date published seventeen books, the cumulative royalties from which do not cover the Council Tax on a small mid-terrace ex-Council house!

But there are other reasons, and hence non-material returns, for communication. Perhaps some are egotistical. However, many more relate to the fact that knowledge and insights that are not communicated in an accessible form are not going to influence the world in which we live. It is that primary motive that leads many of us in the 'green writing' camp to devote the long hours that we do. Another reason is that I will sometimes write to learn. In the depths of early 1990s drought, I was invited to present at a conference convened around the issue of drought impacts. I chose as my title something around the benefits of drought for aquatic life; the ensuing publication 'The importance of periodic droughts for maintaining diversity in the freshwater environment' (1996: Freshwater Forum, 7:2, 33–50) continues to be cited nineteen years on as it took an opposing stance to then dominant ideas about drought and exposed some new thinking.

For many of us 'green scribblers', apart from those landing long-term contracts with better-resourced newspapers and related media or independently wealthy, writing in all its forms is an adjunct to our mainstream academic, NGO, business, government or other spheres of employment or engagement. (I, for one, have never heard anyone utter the phrase, 'The green journalist will join the meeting later when (s)he's found a place to park their Ferrari'!) We write to articulate our thoughts in a coherent and transferrable form, aiming to add to societal discourse and to influence action.

But we can and should protect our interests. You must ensure that you retain copyright on your written and graphic content, though rights to the published form it takes will be transferred to the publisher. The publisher in turn should defend your interests against plagiarism and make agreements on right to your work transferred to other media (magazine serialisation, television/radio, translations, etc.). Photographic images may be of particular potential intellectual property value to you, but that value gets hard to defend once someone – hopefully not you – puts them out into a publicly accessible medium such as an unprotected web page without copyright watermarking. The borderline of what is public property and what is not has got muddier in the digital age.

You also need to be persistent. There is that old saying that 'Everyone has a book in them'. But getting from theory to hard deliverable is more about sweat than inspiration.

Here is a tip for the way I have learned to marshal my writing time, for whatever medium. When I feel creative – sometimes when sat at home, or more often when bored and not tied to immediate pressures in a tedious meeting or on a train journey – I write STRUCTURES rather than prose. This typically takes the form of a bullet list of headers and sub-headers so that I end up with a skeleton structuring what that idea is all about. To this, I can add flesh and blood when I have time but less inspiration. Like many, started off writing like a maniac when inspired but then ran out of time, the flow and destination of the idea then eluding me when I picked up the draft again at a later date, or was inspired by another idea entirely. Most of all, be clear TO WHOM your writing is directed. This sets the style and pitch at which you are writing, the degree of cross-referencing required if any, the balance of polemics versus scientific rootedness, use of first person voice versus factual reporting, 'purpleness' or conversely 'black-and-whiteness' of prose, etc.

Also, sleep on what you have written. The written word is a conscious manifestation, but our subconscious minds do all the creativity. Brilliant ideas may surface unbidden when you wake in the night or are watching TV, so make sure you have a notebook or similar to catch them. And put distance between yourself and drafts you have produced – leave them a day or so – so that you approach them not merely in a less jaded state but also potentially with the fresh eyes of someone encountering your masterwork for the first time; often things that might have been blindingly obvious to you when writing may appear completely opaque when you read them back as an external observer!

Finally, enjoy it! It is, it is true, often a graft to finish a piece of work. But if it's not in you to see it through then it is more productive to do something else more useful and fulfilling for your particular personality and priorities.

As regards other people in this area, well best bet is to look at author names in mags, blogs, etc…I don't think I am better placed that you to do this. It takes a certain type of nutter to want to put their head above the parapet and also to devote much of their life to communications without material reward!

Mark's books include:
The Ecology of Everyday Things (Cultured Llama)
Breathing Space: the Natural and Unnatural History of Air (Zed Books)
River Habitats for Coarse Fish: How Fish Use Rivers and How We Can Help Them (Old Pond Publishing)

Source: Dr Mark Everard, Author, Broadcaster and Associate Professor, University of the West of England, UK. © John Wiley & Sons.

14.9 External Resources

United Kingdom

Conservation Careers – How to be a wildlife journalist
www.conservation-careers.com/conservation-jobs-careers-advice/how-to-be-a-wildlife-journalist/

Mya-Rose 'Birdgirl' Craig
www.birdgirluk.com

Black Girls Hike UK
@UkBgh

Black2Nature
@officialb2n

Black and Green ambassadors
@ujimaBlackGreen

Europe

European Journalism Observatory (EJO) (2020) 'Do European media take climate change
   seriously enough?'
www.en.ejo.ch/specialist-journalism/do-european-media-take-climate-change-seriously-
   enough

European Commission (2020) 'Media Freedom Projects'
www.ec.europa.eu/digital-single-market/en/media-freedom-projects

European Journalism Centre
www.ejc.net

EUSJA
www.eusja.org/about

Asia

Reuters Institute
www.reutersinstitute.politics.ox.ac.uk/risj-review/chinese-environmental-journalism-and-
   sustainable-development

UNESCO
www.en.unesco.org/news/promoting-quality-reporting-environmental-journalism-
   uzbekistan

Konrad-Adenauer-Stiftung (2012) 'Environmental Journalism in Asia-Pacific'
www.kas.de/c/document_library/get_file?uuid=35ed2557-510c-0aaf-c61a-306f17f7a0e2
   &groupId=252038

Africa

Oxpeckers Center for Investigative Environmental Journalism
www.oxpeckers.org

Forum for African Investigative Reporters (FAIR)
www.fairreporters.wordpress.com

Arab Media & Society
www.arabmediasociety.com/environmental-journalism-in-the-uae/

North America

Society for Environmental Journalists
www.sej.org

www.sej.org/library/education-environmental-journalism-programs-and-courses

The National Association of Science Writers
www.nasw.org

NAAEE (2018) 'Science training and environmental journalism today: Effects of science
  journalism training for midcareer professionals'
www.naaee.org/eepro/research/library/science-training-and-environmental

Columbia Journalism Review 'Environmental Journalism? Environmentalism?'
www.archives.cjr.org/behind_the_news/environmental_journalism_envir.php

World Resources Institute
www.wri.org/blog/2013/03/troubling-trends-environmental-journalism

The Knight-Risser Prize for Western Environmental Journalism
www.knightrisser.stanford.edu

John B Oakes Award for Distinguished Environmental Journalism
www.journalism.columbia.edu/oakes

The Baron
www.thebaron.info/news/article/2017/07/12/reuters-wins-environmental-journalism-
  award

Green Writers Press
www.greenwriterspress.com

The Future of Science and Environmental Journalism
www.wilsoncenter.org/event/the-future-science-and-environmental-journalism

Black Girls Hike
@BlackGirlsHike

South America

Pulitzer Center (2018) 'Latin American Media Outlets Call for Increased Climate Reporting'
www.pulitzercenter.org/blog/latin-american-media-outlets-call-increased-climate-
  reporting

EurekAlert! American Association for the Advancement of Science (AAAS) (2020) 'A new
  cross-border science journalism initiative for Latin America'
www.eurekalert.org/pub_releases/2020-02/hhmi-anc021020.php

InquireFirst
www.inquirefirst.org

International Federation of Journalists (IFJ) Latin America
www.ifj.org/where/latin-america-and-the-caribbean/ifj-latin-america.html

Oceania

TheGuardian.com 'Our wide brown land'
www.theguardian.com/environment/series/our-wide-brown-land

Pacific Media Centre
www.pmc.aut.ac.nz/categories/environmental-journalism

Science Media Centre 'Who's reporting science-related issues in New Zealand?'
https://www.sciencemediacentre.co.nz/journos

Global

ICT (2016) 'The 87 Best Blogs for Treehuggers and Climate Change Warriors'
www.ictcompliance.com/blog/the-87-best-blogs-for-treehuggers-and-climate-change-
    warriors/

Greenmatch (2019) 'Top Environmental Bloggers'
www.greenmatch.co.uk/blog/2016/09/top-environmental-influencers-2016

Stephen Leahy
www.stephenleahy.net/

World Federation of Science Journalists
www.wfsj.org

SciDev.Net
www.scidev.net/global/communication/journalism

Association for the Study of Literature and Environment (ASLE)
www.asle.org

Reporters without Borders
www.rsf.org/en/news/environmental-journalism-increasingly-hostile-climate

David Roberts (2013) 'Climate change and environmental journalism'
www.grist.org/climate-energy/climate-change-and-environmental-journalism/

WWF (2009) Environmental journalism and its challenges
wwf.panda.org/?158642%2FEnvironmental-journalism-and-its-challenges

Felix Dodds
www.blog.felixdodds.net/

Minorities in Polar Research
@PolarImpact

Conservation Optimism blog
www.conservationoptimism.org/our-blog

We are all Wonderwomen
www.weareallwonderwomen.com

SEAL (Sustainability, Environmental Achievement & Leadership) Environmental
  Journalism Awards
www.sealawards.com/environmental-journalism-award-winners/

The Guardian UK (2017) 'Guardian wins at the 2017 SEAL Environmental Journalism Awards'
www.theguardian.com/gnm-press-office/2017/oct/02/guardian-wins-at-the-seal-2017-
  environmental-journalism-awards

Earth Journalism Network (EJN)
www.earthjournalism.net

Environmental Journalists
www.environmentaljournalists.org/home

**FOR UPDATED AND ADDITIONAL RESOURCES, INCLUDING EXTRA CHAPTERS, GO TO
WWW.ENV.CAREERS**

References

Columbia Journalism Review (2016). Employment picture darkens for journalists at digital
  outlets. https://www.cjr.org/business_of_news/journalism_jobs_digital_decline.php
  (accessed 14 April 2021).
Environmentalscience.org (2020). What is an environmental writer? www.
  environmentalscience.org/career/environmental-writer (accessed 18 September 2020).
Greenslade, R. and The Guardian UK (2016). Publishers and journalists must work together to
  save journalism. www.theguardian.com/media/greenslade/2016/sep/23/publishers-and-
  journalists-must-work-together-to-save-journalism (accessed 14 April 2021).
International Federation of Journalists (IFJ) (2018). Time to end impunity: IFJ urges drastic
  change in media safety after 82 journalists killed in 2017. https://www.ifj.org/media-centre/
  reports/detail/time-to-end-impunity-ifj-urges-drastic-change-in-media-safety-after-82-
  journalists-killed-in-2017/category/press-releases.html (accessed 14 April 2021).
Press Gazette (2017). How the rise of online ads has prompted a 70 per cent cut in journalist
  numbers at big UK regional dailies. www.pressgazette.co.uk/
  how-the-rise-of-online-ads-has-prompted-a-70-per-cent-cut-in-journalist-numbers-at-big-
  uk-regional-dailies/.

Reuters Institute for the Study of Journalism (2016). Report 'Journalists in the UK'. https://reutersinstitute.politics.ox.ac.uk/our-research/journalists-uk (accessed 14 April 2016).

US Bureau of Labor Statistics (2016). Employment trends in newspaper publishing and other media, 1990–2016. www.bls.gov/opub/ted/2016/employment-trends-in-newspaper-publishing-and-other-media-1990-2016.htm mm(accessed 14 April 2016).

15

Emerging and Other Sectors

There are many other sectors I could have added to this book, probably double the current listings. There are several chapters that could have been divided. In addition, the chapter naming and 'boundaries' could have been organised in different ways.

The Env.Careers website www.env.careers will be the focus for additional content, updates and notes.

There are many emerging and already established sectors in the global environment. The sections below introduce some of these.

NGOs and Charities

There are growing numbers of environmental NGOs globally. At times, they are multinational such as Greenpeace and WWF. Often, they are smaller and more focused in terms of their work programmes.

Traditionally, NGOs are defined by being separate from government organisations in terms of structure and organisation, but often they receive government funding and can also deliver government work programmes. It is often argued that NGOs are a cost-effective route to work delivery.

The NGO sector is very diverse – there are millions of NGOs worldwide, depending on their definition and registration. In many countries, NGOs are informal due to their political sensitivity and pressure from officials and government. They cover environmental, social, political and development issues amongst many others. International NGOs are smaller in number and are often separately defined because they have more formalized structures and significant budgets and staffing levels.

The UN Department of Economic and Social Affairs (DESA) and Economic and Social Council (ECOSOC) have an NGO Branch, which registers NGOs with the UN in terms of offering support.

Consultative status with this unit provides NGOs with access not only to ECOSOC but also to its many subsidiary bodies. As of 2020, 4045 NGOs enjoy consultative status with ECOSOC.

www.un.org/development/desa/dspd/civil-society/ecosoc-status.html

---

Global Environmental Careers: The Worldwide Green Jobs Resource, First Edition. Justin Taberham.
© 2022 John Wiley & Sons Ltd. Published 2022 by John Wiley & Sons Ltd.

Resources

Human Rights Careers '10 human rights organizations offering entry level NGO jobs' www.humanrightscareers.com/magazine/10-human-rights-organizations-offering-entry-level-ngo-jobs

Kimberly Yu, AsianNGO 'Top 5 Benefits of working for an NGO' www.asianngo.org/magazine/post-magazine/article/article-detail/114/top-5-benefits-of-working-for-an-ngo

allaboutcareers.com 'Charity, Not-for-Profit & NGO Careers' www.allaboutcareers.com/careers/industry/charity-not-for-profit-ngo

Jade Phillips, CharityJob (2019) 'What Does It Really Mean to Work For a Charity?' www.charityjob.co.uk/careeradvice/what-it-means-to-work-for-a-charity

Food Sustainability and Regenerative Agriculture

The organisation Sustain notes:
'There is no legal definition of 'sustainable food', although some aspects, such as the terms organic or Fairtrade, are clearly defined. Our working definition for good food is that it should be produced, processed, distributed and disposed of in ways that:
- Contribute to thriving local economies and sustainable livelihoods – both in the UK and, in the case of imported products, in producer countries;
- Protect the diversity of both plants and animals and the welfare of farmed and wild species,
- Avoid damaging or wasting natural resources or contributing to climate change;
- Provide social benefits, such as good quality food, safe and healthy products, and educational opportunities.'
Terra Genesis International define regenerative agriculture as:

> 'a system of farming principles and practices that increases biodiversity, enriches soils, improves watersheds, and enhances ecosystem services. Regenerative Agriculture aims to capture carbon in soil and aboveground biomass, reversing current global trends of atmospheric accumulation. At the same time, it offers increased yields, resilience to climate instability, and higher health and vitality for farming and ranching communities. The system draws from decades of scientific and applied research by the global communities of organic farming, agroecology, Holistic Management, and agroforestry.'

Resources

Spoon University

www.spoonuniversity.com/lifestyle/thoughts-dessert-for-dinner

Audrey Jenkins and Kate Johnson, FoodPrint (2018) 'Get to Work! Jobs in Food Sustainability' www.foodprint.org/blog/get-to-work-jobs-in-food-sustainability

European Commission (2016) 'Sustainable Food'
www.ec.europa.eu/environment/archives/eussd/food.htm#:~:text=For%20food%
2C%20a%20sustainable%20system,such%20as%20climate%20change%2C%
20biodiversity%2C

Sustainable Food Trust
www.sustainablefoodtrust.org

Sustain

www.sustainweb.org/sustainablefood/what_is_sustainable_food

Alexandra Groome and Rachel Kastner, Regeneration International (2017) 'How to
Cultivate a Career in Regenerative Agriculture: Interview with TGI's Ethan Soloviev'
www.regenerationinternational.org/2017/04/04/cultivate-career-regenerative-
agriculture-interview

Terra Genesis International 'Regenerative Agriculture'
www.regenerativeagriculturedefinition.com

Matt Carlson, Training.com.au (2020) 'How to Start a Career in Regenerative Agriculture'
www.training.com.au/ed/regenerative-agriculture

Janet Ranganathan, Richard Waite, Tim Searchinger and Jessica Zionts, World
Resources Institute (2020) 'Regenerative Agriculture: Good for Soil Health, but
Limited Potential to Mitigate Climate Change'
www.wri.org/blog/2020/05/regenerative-agriculture-climate-change

The Circular Economy

The Ellen MacArthur Foundation notes:
'Looking beyond the current take-make-waste extractive industrial model, a circular
economy aims to redefine growth, focusing on positive society-wide benefits. It entails
gradually decoupling economic activity from the consumption of finite resources, and
designing waste out of the system. Underpinned by a transition to renewable energy
sources, the circular model builds economic, natural, and social capital. It is based on
three principles:
• Design out waste and pollution
• Keep products and materials in use
• Regenerate natural systems'
In terms of business and employment opportunities, Elsa Wenzel highlights:
   'The circular economy is celebrated as a trillion-dollar opportunity beginning to pen-
etrate industries around the world. There's no sector or region left untouched by the
potential for reinventing systems, products and services in a fashion that ultimately
creates no waste and even regenerates natural systems. . . Judging by the corporate
momentum toward circularity, the opportunities are only going to expand. Just as jobs
in solar and wind power in the emerging renewable energy landscape outpaced work
in the sputtering coal industry within a mere decade, so too will a circular workforce
replace outmoded roles from high-carbon, high-waste economies.'

Resources

GreenBiz

www.greenbiz.com/collections/circular-economy

Elsa Wenzel, GreenBiz (2019) '5 emerging jobs in the circular economy'
www.greenbiz.com/article/5-emerging-jobs-circular-economy

The Ellen MacArthur Foundation
www.ellenmacarthurfoundation.org/circular-economy/what-is-the-circular-economy
www.ellenmacarthurfoundation.org/circular-economy/concept

UN Jobs

www.unjobs.org/themes/circular-economy

European Commission 'EU Circular Economy Action Plan'
www.ec.europa.eu/environment/circular-economy

Green Alliance and Wrap (2015) 'Employment and the circular economy – Job creation
   in a more resource efficient Britain'
www.wrap.org.uk/sites/files/wrap/Employment%20and%20the%20circular%
   20economy%20summary.pdf

International Development

The organisation NET IMPACT notes:
'International development professionals work to reduce or eliminate poverty in developing countries. Practitioners in this vast field target issues ranging from global health
to emerging market investment opportunities, at scales ranging from village-based
enterprises to country-wide financial and government infrastructures.

   There are a variety of different careers that offer experience with international
development. Some opportunities to consider are positions with internationally-
focused non-governmental organizations (NGOs), the Peace Corps. Multilateral donor
organizations, and partners in non-profit and private sectors.

   With such a wide range of issues that international development focuses on, a wide
array of jobs are available. Some of the job titles that a person interested in international development can pursue include advocacy, communications, consulting, fundraising, policy, and research.'

Resources

NET IMPACT 'The Ultimate Guide for International Development'
www.netimpact.org/careers/international-development/big-picture

John Glenn College of Public Affairs 'Career Opportunities in International
   Development'
www.glenn.osu.edu/career/guides-resources/career-guides/Career%
   20Opportunities%20in%20International%20Development.pdf

Kate Warren, Devex '4 steps to transition to a global development career'
www.devex.com/news/4-steps-to-transition-to-a-global-development-career-89871

Environmental Education

The NAAEE notes:

'Environmental education (EE) is a process that helps individuals, communities, and organizations learn more about the environment, and develop skills and understanding about how to address global challenges. It has the power to transform lives and society. It informs and inspires. It influences attitudes. It motivates action. EE is a key tool in expanding the constituency for the environmental movement and creating healthier and more civically-engaged communities.'

This is a rapidly growing sector globally. There are many job boards and agents covering the sector, and this is a very popular career choice.

Resources

Foundation for Environmental Education (FEE)
www.fee.global

The North American Association for Environmental Education (NAAEE)
www.naaee.org

Facebook Group – Resources for Environmental Education
www.facebook.com/groups/8548813173/permalink/10157575836408174

UK National Association for Environmental Education
www.naee.org.uk

Sustainability and Environmental Education (SEEd)
www.se-ed.co.uk/edu

Forestry and Arboriculture

The National Land Based College offers a helpful definition:
'Forestry and arboriculture are all about trees but offer varied careers. Forestry is the science and practice of planting, managing and harvesting forests for wood and timber, both on a small and large scale. Arboriculture is the cultivation and management of individual trees in a wide range of different environments. There is some overlap in the skills required, but as a forester you would be producing timber on a larger scale and as an arborist you would be looking at maintaining trees in towns, cities, parks, and private gardens.'

The UNECE and FAO highlighted the role of Forestry in green jobs creation:
'Forest experts agree that management of forest ecosystem services is a great way to create green jobs in the forest sector. Until recently, forest jobs were mostly associated with traditional activities related to silviculture (the growing and cultivation of trees) and timber harvesting. Today, the potential for the creation of new job opportunities is enhancing all forest ecosystem functions.'

Tomaselli notes:
'For some South American countries…Forest plantations are a competitive business in the region, and their expansion can immediately increase employment. The cost involved in a plantation establishment programme will most probably be equivalent to

that of any social programme that would be put in place to support unemployed workers. The difference is that while mitigating the social effects of the crisis, the programme would also be creating value.

Several countries in South America, including Brazil, Chile and Uruguay, have already demonstrated the effectiveness of including support for the establishment of large plantations in the national development strategy. These countries are currently the main receivers of direct investments in the forest sector in the region. As a result, in these countries the forest sector is an important contributor to national socio-economic development.'

Key job titles in the sector include forest manager, woodland consultant, forester, silviculture specialist and forest operations supervisor.

Resources

UNECE and FAO (2018) 'Green Jobs in the Forest Sector'
www.unece.org/fileadmin/DAM/timber/publications/DP71_WEB.pdf
www.unece.org/forests/areas-of-work/policy-dialogue-and-advice/green-jobs.html

Tomaselli (2009) FAO 'How forest plantations can contribute to economic renewal in South America'
www.fao.org/docrep/012/i1025e/i1025e06.htm

National Land Based College 'Careers in Forestry & Arboriculture'
www.nlbc.uk/careers/land-based-careers/forestry-arboriculture-careers/?sector=fa#:~:text=Occupations%3A%20Harvesting%20forester%2C%20lumberjack%2C,manager%2C%20silviculturist%2C%20forest%20farmer.

Forestry USA 'Forestry Jobs in America'
www.forestryusa.com/jobs.html

US Forest Service
www.fs.usda.gov/working-with-us/jobs

UN Jobs Forest Management
www.unjobs.org/themes/forest-management

Cyber-Sierra 'Natural Resources Job Search'
www.cyber-sierra.com/nrjobs/forest.html
www.cyber-sierra.com/nrjobs/forest2.html

16

Contributors

Listed below are some of the contributors to this book. There were hundreds of people worldwide who generously passed me thoughts, ideas, comments, case studies, profiles and views; I am sure that some have not been named here. I know that all the contributors, however deeply involved, enriched this book.

Nashon Amollo
Fabiano de Andrade Correa
Rachel Antill
Nick Askew
Tony Barbour
Martin Baxter
James Beal
Chhaya Bhanti
Audrey Boraski
David Brown
Alina Burdulea
Fiona Burns
Maria D. Carvalho
Ruth Chambers
Teresa Conneelly
Adrian Cooper
Gill Cotter
Charly Cox
Janavi Da Silva
Ian Dolben
Kevin Doyle
Alissa Moenting Edwards
David Ellis
John Esson
Mark Everard
Susan Feathers
Veronica Fernandes

Global Environmental Careers: The Worldwide Green Jobs Resource, First Edition. Justin Taberham.
© 2022 John Wiley & Sons Ltd. Published 2022 by John Wiley & Sons Ltd.

Megan Fraser
Kevin Fox
Zoe Greenwood
Abdulhamid Gwaram
Catriona Horey
Shannon Houde
Tiernan Humphrys
Sarah Hussey
Elina Iervolino
Peter Jones
Trish Kenlon
Katie Kross
Phil Le Gouais
Charlotte Lin
Angela Lowe
Kristina Lynn
Joanne Martens
Russell Martin
Carol McClelland
Alun McIntyre
David McRobert
Beth Môrafon
Nadine Bowles Newark
Bosibori M. Ogega
Marita (Ariel) Oosthuizen
Yuly Salazar Páez
Carol Parenzan
Francesca Quell
Fernando Rebelo
Eugenie Regan
Michael Rhodes
Ngaio Richards
Monica Salirwe
Justine Saunders
Hillary Shipps
Penny Simpson
Sharmila Singh
Jamie Sneddon
Alexandros Stefanakis
Laura Thorne
Mike Tregent
Mikael Trewick
Marilyn Waite

Sydney Wang
Claire Wansbury
Sarahjane Widdowson
Stephen Wise
Eleanor Woodhouse
Erika Yarrow
Sarah Yazouri
Lisa Yee-Litzenberg

Index